STP 1177

Rock for Erosion Control

Charles H. McElroy and David A. Lienhart, editors

ASTM Publication Code Number (PCN)
04-011770-38

ASTM
1916 Race Street
Philadelphia, PA 19103

Library of Congress Cataloging-in-Publication Data

Rock for erosion control / Charles H. McElroy and David A. Lienhart,
    editors.
        (ASTM special technical publication (STP) ; 1177)
        Includes bibliographical references and index.
        ISBN 0-8031-1489-3
        1. Rockfills.   2. Soil stabilization.   3. Soil conservation.
    I. McElroy, Charles H.   II. Lienhart, David A.   III. Series.
    TA709.R63   1993
    624.1'51363—dc20   92-44625
                                                              CIP

Photocopy Rights

Peer Review Policy

Each paper published in this volume was evaluated by three peer reviewers. The authors
addressed all of the reviewers' comments to the satisfaction of both the technical editor(s) and the
ASTM Committee on Publications.

The quality of the papers in this publication reflects not only the obvious efforts of the authors and
the technical editor(s), but also the work of these peer reviewers. The ASTM Committee on
Publications acknowledges with appreciation their dedication and contribution to time and effort on
behalf of ASTM.

Printed in Philadelphia, PA
January 1993

Foreword

This publication, *Rock for Erosion Control*, contains papers presented at the symposium, Durability and Specification Conformance Testing of Rock Used for Erosion Control, held in Louisville, KY on 18 June, 1992. The symposium was sponsored by ASTM Committee D-18 on Soil and Rock and its Subcommittee D18.17 on Rock for Erosion Control. Charles H. McElroy of the Soil Conservation Service in Forth Worth, TX and David A. Lienhart of the U.S. Army Corps of Engineers in Cincinnati, OH presided as symposium co-chairmen and are the editors of the resulting publication.

Contents

Overview

Rock (stone) has been widely used for many years as a means to combat erosion, especially that caused by hydraulic forces. Millions of tons of quarried stone are used each year in the United States and throughout the world. Because a wide variety of rock types and sources are available, a multitude of uses are involved, and exposure to climatic conditions vary widely, no concensus on criteria exists among those who must evaluate the durability of the rock for the proposed design life of the associated structures.

This Special Technical Publication (STP) has been published as a result of the June 1992 symposium on Durability and Specification Conformance Testing of Rock Used for Erosion Control held in Louisville, Kentucky. The symposium was an outgrowth of work within ASTM Subcommittee D18.17 on Rock for Erosion Control, a subcommittee of ASTM Committee D18 on Soil and Rock.

ASTM Subcommittee D18.17 was formed in 1878 as a result of expressed needs by those in attendance at a special Task Group meeting held in Boston, Massachusetts. The approved scope of D18.17 is as follows:

> It shall be the responsibility of Subcommittee D18.17 to develop methods of tests and engineering specifications for rock (variously known as riprap, breakwater stone, armor stone, filter stone, and bedding material) for erosion control.

Several needs were identified by Subcommittee D18.17. Generally these needs fall into one of three categories: (1) rock durability testing, (2) rock gradations (specifications), and (3) acceptance testing.

An informal survey of some of the major users of rock for erosion control in the United States indicated that the most common laboratory tests used for evaluating durability of rock were petrographic examination, wet-dry, freeze-thaw, sodium or magnesium sulfate soundness, specific gravity, absorption, Los Angeles Abrasion, and the splitting tensile strength. The test procedures used varied from user to user, but were generally very similar to existing ASTM standard test methods for coarse aggregates. As a result, very small test specimens were and still are generally used. Considerable variation exists among the users on the acceptable criteria for each test. In most cases the criteria are based on the intended use of the rock being tested.

Many designers consider the use of small specimens to be a major limitation to the present state-of-the-art for rock durability testing. Their contention is that the breaking of a large piece of stone into smaller test specimens results in breakage along planes of weakness in the stone. This may, in some cases, indicate a false evaluation of the rock's durability.

Testing of small specimens has evolved to its present state for several reasons:

1. Standard test procedures are already available, and some performance data are documented.
2. Equipment to test larger specimens is expensive and cumbersome to operate.
3. Large scale testing is labor intensive.
4. Documented performance data that can be correlated back to large scale durability testing are lacking.

The second major category of problems identified concerns design specifications. Because a wide variety of uses and many different users exist, designers often have their own unique design procedure to size the rock for their particular application. Complicated, and in many cases very confusing, specifications often result. Some specifications are based on mass and some on size. Generally, a range is given. Often, quarry operations, including equipment, must be changed to meet the specifications, resulting in a much higher unit cost. Some quarries simply refuse to produce a limited tonnage of rock if a change in operations or equipment is required. In addition, many designers do not understand quarry operations, and their specifications are often vague and impractical to meet.

A third prolem area concerns the methodologies used to check for conformance to design specifications. Gradations or sizes can be specified to be checked at the quarry, as delivered to the project site, or after placement. The rock may meet the specifications at the quarry, but hauling and placement can alter the inplace gradations.

Not only is "when" to check gradation a problem, but "how" is perhaps a bigger one. Some inspectors have a truck dump a load of rock on a concrete pad and weigh or measure each stone. Some may just measure the stone, and others use a visual inspection only.

ASTM Committee D18, through the activities of its Subcommittee D18.17, has been developing standards to address the identified problems. ASTM D4992, Standard Practice for Evaluation of Rock to be Used for Erosion Control, was the first standard developed. Others include ASTM D5121, Standard Practice for Preparation of Rock Slabs for Durability Testing, and D5240, Standard Test Method for Testing Rock Slabs to Evaluate Soundness of Rock Riprap by Use of Sodium and Magnesium Sulfate. Standard test methods for wet-dry and freeze-thaw testing of rock slabs are in the ASTM ballot process.

Subcommittee D18.17 has also been developing a proposed specification for sizes of rock used for erosion control. The proposed standard has undergone several drafts and is also in the balloting process.

Members of Subcommittee D18.17 received approval to plan and sponsor a symposium. The goals of this symposium were:

1. To provide an opportunity to examine the current technology used by scientists, geologists, engineers, and others to measure and evaluate the durability and performance of rock used for riprap, gabions, canal and channel linings, and other erosion control applications.
2. To provide a forum for presenting state-of-the-art procedures used by construction and quarry personnel for specification conformance and compliance testing.
3. To identify areas where new standards are needed. The organizers also hoped that the symposium would publicize the work being done in the subcommittee and generate additional assistance in the development of standards already in progress and others that are needed.

The symposium was organized into two sessions: (I) Durability Testing, and (II) Specification Conformance Testing. It featured invited authors and selected authors who responded to the Call for Papers.

The organizers invited a larger user (U.S. Bureau of Reclamation) to present an overview of the procedures their agency uses to evaluate the durability of rock. They also invited a representative from a large quarry operation to present a paper on the equipment, procedures, and problems involved in producing stone for erosion control projects. The third invited participant presented a paper that contained background information and a draft of the proposed standard specification for sizes of rock used for erosion control that was developed in Subcommittee D18.17.

This STP will serve as an excellent reference on both durability and conformance testing of rock for those actively engaged in production, testing, design, and Quality Assurance/Quality Control (QA/QC) activities. Many interesting topics for research have been identified that will challenge and direct our future work in these areas. In part I of the STP, new test methods, such as the mill abrasion test for wear resistance, were introduced. Refinements, improvements, and additional data are also presented for older test methods, such as the jar slake test, insoluble residue test, wet-dry test, and the freeze-thaw test. Another paper introduced the use of petroglyphs and Indian Rock Art age dating as indicators of durability. A mathematical analysis of fractals and pore potential to explain, predict, and document durability was also presented.

Relying on a single laboratory test to evaluate the durability of rock is impossible. One paper presents an excellent summary of the importance of visual inspection and observation and historical performance documentation in evaluating rock for a specific use.

In Part II an invited paper outlines the processes used at a major quarry to produce different size stones to meet specifications. Another invited paper gives the details on a proposed standard specification for sizes of rock. ASTM Subcommittee D18.17 is very interested in any feedback concerning these proposed specifications. The results of a survey of the specifications used by State highway departments and two federal agencies are presented in another paper. In addition, criteria for several different quantitative tests are shown along with generic qualitative requirements listed by some users. Another paper focuses on the improvements needed in present QC/QA methods and verification along with three case histories as examples. Discussions of correlation data based on field performance further enhance the value of the symposium and this publication.

Although each paper presents specific conclusions, some generalized conclusions common to several of the papers are:

- The size of the rock pores is important if rock is subjected to extremes in temperature. Rocks that have small pores are generally more susceptible to damage from climatic changes.
- Although accelerated weathering tests are difficult to model in the laboratory, good correlation to performance has been achieved, especially with the freeze-thaw test.
- Petrographic analysis is a vital element in evaluating durability of rock.
- In addition to laboratory evaluation, visual observations and historical performance data are valuable tools for evaluating rock sources.

Perhaps a more important result of the symposium is that many issues were identified and additional research and work are urgently needed. Some of these are being worked on by ASTM Subcommittee D18.17 and others. However, many of the issues are not being addressed. Readers of the STP will discover several different avenues of research and study that are needed to advance the state-of-the-art technology. Some of the more pressing issues are:

- Consensus standard specifications for sizes of rock used for erosion control.
- Better and easier methods for modeling accelerated weathering in the laboratory.
- An evaluation of the size and type of rock test specimens and how well the test specimens represent actual rock masses (cube vs. slab vs. rock cores vs. broken pieces).
- Improved standard test methods for QA/QC at the source and at the project.

The editors express their appreciation to all those who attended the symposium, the authors whose papers appear in this volume, the reviewers of the submitted papers, the ASTM Committee D18 for sponsoring the symposium through its Subcommittee D18.17, and to the ASTM Staff. Without their combined efforts, this STP would not have reached the high quality level that is associated with ASTM publications.

Charles H. McElroy
Soil Conservation Service
Fort Worth, TX

David A. Lienhart
U.S. Army Corps of Engineers
Cincinnati, OH

Durability Testing

Jeffrey A. Farrar [1]

BUREAU OF RECLAMATION EXPERIENCE IN TESTING OF RIPRAP FOR EROSION CONTROL OF EMBANKMENT DAMS

REFERENCE: Farrar, J.A., "Bureau of Reclamation Experience in Testing of Riprap for Erosion Control of Embankment Dams," Rock for Erosion Control, ASTM STP 1177, Charles H. McElroy and David A. Lienhart, Eds, American Society for Testing and Materials, Philadelphia, 1993.

ABSTRACT: The Bureau of Reclamation has accumulated significant experience with the use of riprap for erosion control of embankment dams through its history as a major water resources design and construction agency for irrigation projects in the seventeen western states. Successful exploration, design, testing,and construction methodologies have been developed through experience. Exploration and design aspects are well documented in a series of technical manuals and design standards. Riprap quality evaluations depend heavily on expert geologic and petrographic evaluations coupled with physical properties and freeze thaw testing of rock specimens. Quality evaluation methodologies were heavily influenced by concrete technology testing resulting in the use of 75 mm (3 in) cube specimens for freeze thaw testing. Physical properties tests are performed on crushed coarse aggregate gradations. Reclamation has an available database of almost 1000 quality evaluations. Field placement, control test procedures, and riprap performance studies are reviewed.

KEYWORDS: Riprap, quality evaluations, exploration, design, construction, control testing, performance

INTRODUCTION

The purpose of this paper is to present Bureau of Reclamation experience with the use of riprap for erosion control. Reclamation, as a major water resources agency in the seventeen western states, has developed proven procedures for exploration, design, quality evaluation, construction and construction control testing through long experience. Much of this experience is documented in numerous technical manuals and design standards cited in this paper.

Data compiled in Reclamation's research laboratory provide a valuable resource of information on rock quality in the western US. Consistent physical properties testing procedures have evolved and remained constant for the last thirty years. These tests along with highly skilled geologic and petrographic determinations allow Reclamation engineers to evaluate rock sources with confidence. Available data also provide information on use of examined quarries as concrete aggregate sources. An additional database of concrete aggregate sources is also available.

---

[1]Supervisory Civil Engineer, Materials Engineering Branch, Research and Laboratory Services Division, Bureau of Reclamation, PO Box 25007 D-3733, Denver, Colorado, 80225

In this paper, a brief review will be given to all aspects of the use of riprap for erosion control.

FIELD EXPLORATION REQUIREMENTS

Numerous guides, such as the Design of Small Dams and Engineering Geology manuals, are available in Reclamation for exploration of rock sources for use as riprap (Bureau of Reclamation 1977, 1988). These references provide information on geologic considerations for evaluating riprap sources. All sources considered for final design are geologically mapped in great detail. Important data include major joint patterns and block development along with detail given to the weak inclusions present. Talus slopes which at first appear to be a good source of riprap, may not be suitable due to weathering or insufficient quantities. For final designs, it is often necessary to perform blast tests generating 7 to 14 m^3 (10 to 20 yd^3) to fully evaluate the deposit. Recommended procedures for blasting are given in the Engineering Geology Manual (Bureau of Reclamation, 1988).

The second edition of the Earth Manual contains an investigation guide, "Investigation of Rock Sources for Riprap," under designation E-39 (Bureau of Reclamation, 1974). This guide was not included in the third edition, part II, of the Earth Manual, but is still used as a guide (Bureau of Reclamation, 1990). We intend to include this guide in the Rock Manual currently under preparation. The guide provides information on sampling, testing, and reporting requirements. This guide has been rewritten recently to reflect current concrete and concrete materials testing procedures soon to be published in the new Concrete Manual, Part II, Ninth Edition. Concrete Manual procedures will be published in 1992.

For laboratory test samples, we depend on good field geology for selection of representative rock specimens. Prior to sampling, detailed geologic study is performed using core drilling, excavation, and even blast tests. In many cases several samples representing the range of quality are obtained. Samples to be sent to Denver Office laboratories are required to have a minimum mass of 273 kg (600 lb) although quite often larger samples are received. The minimum rock particle size in the sample is 0.014 m^3 (1/2 ft^3).

DESIGN CONSIDERATIONS

General considerations for design of slope protection and typical specifications requirements for small dams are reported in the manual Design of Small Dams (Bureau of Reclamation, 1977). Of all slope protection methods considered by Reclamation, use of dumped riprap is by far the most desirable method of slope protection for embankment dams. Effective riprap design must consider the following factors;

- Quality of the rock.
- Weight or size of the individual pieces.
- Thickness of the riprap.
- Shape of the stones or rock fragments.
- Slopes of the embankment on which the riprap is placed.
- Stability and effectiveness of the filter on which the riprap is placed.

Weight or rock sizes are designed based on wave action which must be resisted. Gradations usually found suitable on 3 horizontal to 1 vertical slopes for reservoirs with effective fetches less than and greater than 4.0 km (2.5 miles) are summarized in Table 1. These general guidelines are based on Reclamation experience.

TABLE 1 -- Thickness and gradation limits of riprap on 3:1 slopes
(Bureau of Reclamation, 1977).

Reservoir Fetch km	Nominal Thickness m	Gradation, Percentage of Rock of various weights - kg (1)			
		Maximum Size kg	40 to 50 percent greater than kg	50 to 60 percent greater than kg	0 to 10 percent less than kg (2)
4.0 or less	0.76	1 134	567	34 - 567	34
More than 4.0	0.9	2 041	1020	45 -1020	45

(1) Sand and rock dust shall be less than 5 percent, by mass, of the
total riprap material.
(2) The percentage of this size material shall not exceed an amount
which will fill the voids in larger rock.

Reclamation's detailed design procedures for evaluating riprap
slope protection are well documented in a series of Design Standards
developed by the Geotechnical Engineering and Embankment Dams Branch of
Reclamation. These Design Standards are under varying stages of
development and subject to continual revisions as state of the art
information is obtained. Embankment freeboard design depends on slope
and roughness considerations (Bureau of Reclamation, 1984).
Increasingly, the use of rock fragments for slope protection is
considered unacceptable due to cost considerations or environmental
restraints. In these cases, the use of soil cement slope protection is
considered. Design Standards have also been developed for soil cement
(Bureau of Reclamation, 1991a).

The primary Design Standard for riprap slope protection evaluates
design factors to be considered and design procedures for sizing riprap
and bedding (Bureau of Reclamation, 1990). Some important design
factors include;

    - Wave characteristics
    - Reservoir operations
    - Embankment design
    - Rock quality
    - Others, debris damage, extreme freeze thaw, availability of
    rock, and remote reservoir site.

Design procedures begin by evaluation of design winds and
development of design wave height. Riprap weight is determined by
evaluating forces acting on the rocks and resistance forces offered by
the riprap matrix. Empirical relationships are used for evaluating
stability. These methods were developed by the US Army Corps of
Engineers and further modified with additional research. Factors
influencing the stability of riprap are listed below;

    - Wind velocity, duration, and direction
    - Wave height
    - Wave period
    - Reservoir shape
    - Reservoir depth

- Reservoir fetch
- Direction of wave attack
- Manner in which waves impinge on the embankment (breaker type)
- Number of waves striking the embankment
- Embankment slope
- Roughness of riprap surface
- Porosity of the riprap layer
- Rock particle weight, dimensions, and shape
- Density of rock
- Keying of rock particles
- Thickness of riprap layer
- Support provided by bedding
- Gradation of bedding
- Thickness of bedding

After riprap size range and thickness are determined, the gradation for bedding materials is evaluated by applying filter criteria.

Review of all existing standards indicates that 0.9 m (36 in) riprap is the typical maximum specified by Reclamation although there may be some exceptions in extreme cases. The 0.9 m (36 in) riprap consists of 0.9 m (36 in) maximum size rock fragments placed in a 0.9 m (3 ft) thickness normal to the slope. The 0.9 m (36 in) rock blanket protection along with 0.46 to 0.38 m (18 to 15 in) bedding is considered to protect for most extreme events. The maximum rock fragment size of 0.76 m^3 (1 yd^3) is considered to be the maximum size available from typical quarry operations. Larger sizes may be available from some rock sources but consideration must be given to construction equipment required for handling.

QUALITY EVALUATIONS

Quality evaluations are performed in the Denver Office laboratories using procedures developed in the mid 1900's. Essentially, the same procedures have been used for the last 30 years. The Denver Office laboratory has almost 1000 records of quality testing of riprap samples using procedures cited below. When the same source is also considered a source for concrete aggregates, additional testing is performed. Additional test data for concrete evaluations include performance of concrete specimens cast in a standard proportioning for strength and freeze thaw resistance. The riprap and concrete aggregate data available in the Denver Laboratory represent a valuable resource for rock quality data in the western United States. Records are maintained by the Materials Engineering Branch, Research and Laboratory Services Division and are available upon request. These data could provide useful research information into quality evaluation tests if incorporated into a database. Proposals have been made to compile this database into computer form but funding sources have not been found. Other agencies or organizations are welcome to access these data for cooperative research.

Reclamations test procedure development was intimately related to developing concrete technology. Based on development of the freezing and thawing apparatuses for 75 mm (3 in) diameter concrete test cylinders, the use of 75 mm (3 in) freeze thaw cubes has been standardized. Also, since many sources provide for combined use as concrete aggregates, physical properties tests consisting of specific gravity, absorption, Los Angeles abrasion, and sodium sulfate soundness are performed on crushed materials representing coarse concrete aggregate gradations.

Quality evaluation testing is heavily dependent on petrographic examinations in addition to properties testing. Guidelines for petrographic examinations are given in a petrographic laboratory manual and the test procedure for petrographic examination of concrete aggregates (Sheldon, 1985, Bureau of Reclamation, In Preparation). It is most important to maintain a staff of skilled petrographers who are familiar with the use of riprap. Petrographic examinations may reveal important defects in rock which may seem satisfactory from laboratory tests.

Standard laboratory tests performed on rock specimens include;

    Freeze thaw durability testing of 75 mm (3 in) cubes

    Physical properties of coarse aggregate grading consisting of;
        Specific gravity
        Absorption
        Sodium sulfate soundness
        L.A. abrasion resistance

Freeze thaw tests serve to model the performance of rock in cold climates and constitute an important criterion for acceptance in both cold and warm areas. Freeze thaw tests are performed on 75 mm (3 in) rock cubes cut from rock fragments selected to represent the poorest, medium, and best quality rock on the basis of visual examination by the petrographer. Cubes are preferred over slabs since all 6 faces of the cube have smooth surfaces. These control surfaces allow for location of failure and result in better understanding of causes of failures. The failures are closely examined by petrographers.

Freeze thaw cubes are inserted into 3 inch rubber sheaths and sufficient water is added to cover the specimen. At Reclamation's test facility, cube samples are subjected to 50 cycles of freezing and thawing every week. Each cycle consists of 1-1/2 hours freezing at -12 degrees C (10 degrees F) and 1-1/2 hours thawing at 21 degrees C (70 degrees F). Thermocouple studies of 75 mm (3 in) diameter concrete specimens have shown that these cycle periods are long enough for complete freezing and thawing. The test is continued for 250 cycles or until 25 percent mass loss occurs. Mass loss is computed by comparing initial cube mass to the mass of the largest remaining fragment. Cube specimen failure modes (e.g. fracture, joint, or bedding plane split; disaggregation; exfoliation) are noted and related to riprap rock durability requirements.

Physical properties tests are performed on material crushed from rock fragments to meet coarse concrete aggregate gradation. This is a compromise because the crushing process tends to remove structural defects in the rock. The only benefit to crushing is when the rock source is considered for concrete production because the data on crushed aggregate will satisfy both riprap and concrete aggregates testing needs. The crushing process tends to reflect the performance of rock without structural defects, that is, the background mineralogy. Reclamation's procedures are very similar to those of ASTM. Table 2 compares USBR Concrete Manual procedures to ASTM procedures (ASTM 1992a, 1992b, Bureau of Reclamation, In Preparation).

After testing is complete, the petrographer and aggregate specialist review available data and make recommendations on acceptability for intended uses. Table 3 provides general criteria in terms of test results that are considered typical of acceptable riprap. No single test has proven superior to evaluate rock quality. The results of any single test should not be used as the sole basis for approval of sources. Petrographic examinations have the most weight. Petrographers and aggregate specialists indicate that there is a correlation between sodium sulfate soundness and freeze thaw losses. This may be a possible area of future study of Reclamations database.

TABLE 2 -- Comparison of USBR and ASTM testing procedures for riprap.

Test	USBR PROCEDURE (CONCRETE MANUAL)	ASTM PROCEDURE
Freezing and Thawing	4666-90	C 666
Sodium Sulfate Soundness	4088-89	C 88
Specific Gravity, Absorption (Coarse)	4127-90	C 127
Los Angeles Abrasion (Coarse)	4131-90	C 131
Petrographic Examinations	4295-89	C 295
Sampling and Testing (Old: Earth Manual E-39)	7xxx-90 (Rock Manual)	D 4992
Sample Reduction (Coarse Aggregate)	4702-90	C 33, 702

TABLE 3 -- Typical acceptable properties of riprap.

TEST	ACCEPTABLE RANGE
Freeze Thaw Durability	< 25% loss @ 250 cycles (1)
Specific Gravity	> 2.60
Absorption	< 2%
Sodium Sulphate Soundness	< 10% loss (2)
Los Angeles Abrasion	< 10% loss @ 100 cycles < 40% loss @ 500 cycles

(1) Freeze thaw loss criteria depend on petrographic examination and judgement of the modes of failure.
(2) <15% loss for crushed limestone is acceptable.

FIELD PLACEMENT AND CONTROL

        Specifications for riprap construction vary depending on scope of the project. Typical specifications paragraphs for embankment dams are given in the Design of Small Dams manual (Bureau of Reclamation, 1977). Provisions are made for quarry operation, quality (composition and shape), and placement. In many cases, approved sources are identified in advance. If a different source is proposed, quality testing is specified.

        Typical specifications wording for composition and shape are as follows:

"Riprap Composition - Riprap shall consist of various sizes of hard, durable, and sound rock obtained by selective quarrying, processing of quarried rock, and/or processing of rock obtained from rock borrow sites. The materials shall not contain boulders or cobbles from surficial deposits, organic materials, roots, debris, or other deleterious material. All rock used for riprap shall be of (*insert allowable rock types*) as defined and shown on the drawings as (*insert allowable geologic symbols*). Acceptable riprap shall be free from open or incipient cracks, seams, structural planes of weakness, or other defects that would tend to increase unduly its deterioration from natural causes and from handling and placing.

"Riprap Shape - Riprap shall be predominantly angular and blocky in shape rather than elongated as more nearly cubical stones nest together best and are more resistant to movement. The riprap shall have sharp, clean edges at the intersection of relatively flat faces. The following shape limitation is specified for riprap obtained from quarry operations:

(1) Not more than 25 percent by weight of the riprap reasonably well distributed throughout gradation shall have a length more that 2.5 times breadth or thickness.

(2) Elongation tests will be run on the plus 150 mm (6 in) material only. These limitations apply only to the rock within the riprap gradation."

Specifications for riprap grading have varied widely depending on needs of the project. Specifications which provide a detailed range for allowable rock sizes or weights at a single percentage passing are preferred to provide clarity to the contractor. Some specifications have ranges for both rock fragment size and percent passing or retained. The use of dual ranges is discouraged because it adds some confusion to the exact desired product.

An example of specifications requirements from a riverbank stabilization project are shown on Table 4 and Figure 1. On this project, gradation tests were performed by weighing particles. A specification weight range is given for eight percentage passing levels. This is easily depicted on Figure 1 showing riprap stone size distribution. The only additional requirement is that the gradation curve be parallel to the boundary lines to obtain a good representation of all rock sizes within the range.

TABLE 4 -- Example gradation requirement for riprap.

Percentage finer by weight	Approximate individual stone weights	
	Maximum weight - kg (lb)	Minimum weight - kg (lb)
100	9 370 (4 250)	3 970 (1 800)
90	7 940 (3 600)	3 300 (1 500)
70	5 510 (2 500)	2 200 (1 000)
50	3 970 (1 800)	1 540 (700)
30	1 540 (700)	485 (220)
15	770 (350)	200 (90)
5	485 (220)	110 (50)
<5	110 (50)	

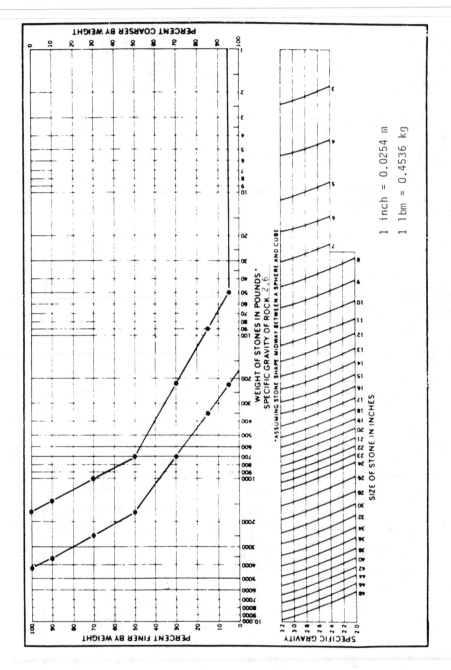

FIGURE 1 -- Example gradation requirement for riprap.

Placement methods have evolved with the advent of new equipment. The first consideration in placement is that the quarry operations are such that a good mixture of rock sizes is available in each load delivered to the site. Gradation tests are performed on loads before placement to ensure that size distribution requirements are met. Placing of the loads should be performed to ensure that segregation does not occur. Placement is accomplished by placing loads along the slope against previously placed riprap to prevent the segregation that occurs if dumped in piles. Dumping from the top of a slope into a chute is not allowed. Dumping should proceed in horizontal rows and progress up slope. In recent years backhoes with 1.1 to 1.9 m^3 (1.5 to 2.5 yd^3) capacity buckets have become the most common method for placement (Bureau of Reclamation, 1991b). With backhoes, the riprap must be kept close to embankment level for the arm to reach below the slope. Other successful methods for placing riprap include dragline with skip, Grade Alls, cranes with clamshells, and rubber tired front end loaders. Continual use of visual inspection is required during placement to ensure proper mixing and interlocking of the rock fragments. In most cases some reworking by hand is required but it can be minimized by proper loading and placement.

Gradation testing has varied widely depending on success of quarrying operations. On minor projects testing is based only on less reliable visual methods such as photogrids. Gradation tests are often specified on larger projects to check early operations and periodic monitoring as operations require.

Currently, Reclamation has no test procedure or guide for determining the particle size distribution for riprap. We plan to add a procedure to Reclamation's Rock Manual after methods are developed and approved by ASTM . Gradation tests can be performed using volumetric or mass measurement methods. Recently, the preference is to use mass measurement methods of individual rock fragments with load cells. New hanging type load cells can be used with a clamshell or orange peel for weighing individual rock fragments. Riprap stones greater than 23 kg (50 lb) are weighed individually while others less than 23 kg (50 lb) can be weighed together. Nine by 9 m (30 by 30 ft) concrete test pads are required as a working surface to process 50 ton samples. Fifty tons is the preferred sample size for 0.9 m (36 in) riprap similar in gradation to the specification example above. If riprap sizes need to be evaluated from particle mass, the size can be estimated assuming stone shape between a sphere and a cube as shown on Figure 1 from the specification example above.

There have been many differing techniques for gradations based on volumetric measurements. Methods are usually developed by field laboratory chiefs. Volumetric methods start by grouping rock fragments into size ranges that may be specified. In some cases, only visual methods are used to sort rock fragments. In other cases templates are used for coding size ranges. Rock fragments are often marked and grouped for weighing.

PERFORMANCE OF RECLAMATION DAMS

For this paper, it was not possible to perform additional studies of Riprap performance or laboratory correlations from Reclamation's database. There are however, previous studies of riprap performance worth reviewing.

In 1968, the Dams Branch performed a study of 149 existing Reclamation dams. This study, conducted by E. Esmiol, had major influence in the use of Riprap on embankment dams (Esmiol, 1967). In the report the following items were examined; condition of upstream slopes, survey of rock types, causes of apparent failures, cost trends, and features of excellent riprap such as gradation, specifications, design, laboratory testing, field explorations, and quarry sources. Unfortunately, the dams studied were of older construction and relevant evaluation of existing testing procedures developed in the 1950's and 60's could not be made. The report includes information on all major rock types and some unusual rock types such as chalk and alluvial cobbles and boulders.

Of the 149 dams surveyed, 56 were judged in excellent condition, 27 in good condition, 11 in satisfactory condition and 4 were not classified. A total of 51 dams were considered to have failed. Failure in this report was considered to be a failure to perform up to expectations. No riprap failures resulted in embankment dam failure. Failures could be grouped into seven categories as follows;

1. Change of shape (settlement, sloughing, or subsidence) due wind and wave action or poor construction.
2. Erosion failure caused by abrasion, beaching, washing of fines, displacement, and plucking of stones due to wind, wave or ice action.
3. Rock breakdown caused by weathering and disintegration due to temperature variations, water absorption, ice action, excessive heat and root actions.
4. Stone removal by others
5. Nonuniform placement
6. Combinations of erosion and change of shape
7. Unknown

There were sixteen failures due to rock breakdown and many of these lacked appropriate field investigations and laboratory testing. Some examples of failures were cases where rock of questionable durability was used intentionally. The report pointed to the need for better field and laboratory evaluations. One example of Keene Creek dam showed that a lack of petrographic examinations was important since all of the laboratory data such as freeze thaw and aggregate quality evaluations had suitable results. A petrographic examination was made on an unrepresentative sample and detected interstitial chlorite material and warned of possible contamination with weathered rock. Apparently at Keene Creek, weathered materials were placed without further petrographic and laboratory testing, resulting in poor performance.

In response to Esmiol's study, a team of Reclamation technical experts was formed to review this report and existing Reclamation procedures to make recommendations for future design and construction (Davis et.al., 1973). Their finding indicated that many of the rock breakdown failures could have been predicted using Reclamation quality evaluations. They stressed the use of petrographic examinations and physical properties tests.

There have been several studies on the failure of riprap on Cedar Bluff Dam due to extreme wind-induced waves. Waves of up to 8 ft were reported. In the most recent reanalysis of the event, it was shown that existing riprap were undersized to handle that particular event (Bureau of Reclamation, 1991a). Adequate protection could have been provided using current design procedures which would specify maximum size rock fragments from 1360 to 2270 kg (3000 to 5000 lb).

CONCLUSION

Bureau of Reclamation practice for the use of riprap for erosion control of embankment dams has been reviewed. Exploration, design, testing, construction and performance aspects have been presented. Practice for exploration and design are well documented in a series of manuals and Design Standards cited in this report. Excellent guidance documents are available to designers. Reclamation test procedures have remained constant for many years and a large database for rock quarries is

available for the Western States. Designers and laboratory staff are
confident in existing procedures. In construction, there is not an
existing Reclamation procedure for grading riprap and many methods have
been used. Additional performance studies of existing riprap
installations are needed.

REFERENCES

American Society for Testing and Materials, 1992a, Annual Book of ASTM
    Standards, Section 4, Construction, Volume 04.08, Soil and Rock;
    Dimension Stone; Geosynthetics, 1916 Race Street, Philadelphia, PA
    19103, USA.

American Society for Testing and Materials, 1992b, Annual Book of ASTM
    Standards, Section 4, Construction, Volume 04.02, Concrete and
    Aggregates, 1916 Race Street, Philadelphia, PA 19103, USA.

Bureau of Reclamation, 1974, "Investigation of Rock Sources for
    Riprap," Designation E-39, Earth Manual, Second Edition, US
    Department of the Interior, Bureau of Reclamation.

Bureau of Reclamation, 1977, Design of Small Dams, US Department of the
    Interior, Denver Colorado, Revised Reprint.

Bureau of Reclamation, September 1984, "Draft Design Standards - No 13 -
    Embankment Dams - Chapter 6 - Freeboard," United States Department
    of the Interior, Denver Office, Denver, Colorado.

Bureau of Reclamation, 1988, Engineering Geology Manual, , US Department
    of the Interior, Denver, Colorado, First Edition.

Bureau of Reclamation, 1990, Earth Manual, Part II, Third Edition, US
    Department of the Interior, Denver, Colorado, Third Edition, 1990.

Bureau of Reclamation, October 1990, "Draft Design Standards - No 13 -
    Embankment Dams - Chapter 7 - Riprap Slope Protection," United
    States Department of the Interior, Denver Office, Denver, Colorado.

Bureau of Reclamation, July 1991a, "Design Standards - No 13 -
    Embankment Dams - Chapter 17 - "Soil Cement Slope Protection,"
    United States Department of the Interior, Denver Office, Denver,
    Colorado.

Bureau of Reclamation, July 1991b, " Design Standards - No 13 -
    Embankment Dams - Chapter 10 - Embankment Construction," United
    States Department of the Interior, Engineering and Research Center,
    Denver, Colorado.

Bureau of Reclamation, In Preparation, Rock Manual, First Edition,
    US Department of the Interior, Denver, Colorado.

Bureau of Reclamation, In Preparation, Concrete Manual, Part II, Test
    Procedures, Ninth Edition, US Department of the Interior, Denver,
    Colorado.

Davis, F.J., Burton, L.R., Crosby, A.B., Klein, L.D., and E.R.
    Lewandowski, March 1973, "Riprap Slope Protection for Earth Dams: A
    Review of Practices and Procedures," Report No. REC-ERC-73-4,
    Engineering and Research Center, United States Department of the
    Interior, Bureau of Reclamation Denver, Colorado.

Esmiol, E.E., September, 1967, Revised 1968, "Rock as Upstream Slope
    Protection for Earth Dams -- 149 Case Histories," Report No. DD-3,
    Dams Branch, Division of Design, Office of Chief Engineer, United

States Department of the Interior, Bureau of Reclamation Denver, Colorado.

Sheldon, G.J., September 1985, "Petrographic Laboratory Analytical Techniques and Capabilities Reference," United States Department of the Interior, Bureau of Reclamation, Denver Office, Denver, Colorado.

Luis E. Vallejo[1], Robert A. Welsh, Jr.[2], C. William Lovell[3], and
Michael K. Robinson[4]

THE INFLUENCE OF FABRIC AND COMPOSITION ON THE DURABILITY
OF APPALACHIAN SHALES

REFERENCE: Vallejo, L. E., Welsh, R. A., Jr., Lovell, C. W., and
Robinson, M. K., **"The Influence of Fabric and Composition on the
Durability of Appalachian Shales,"** Rock for Erosion Control, ASTM
STP 1177, Charles H. McElroy and David A. Lienhart, Eds., American
Society for Testing and Materials, Philadelphia, 1993.

ABSTRACT: The durability of sixty-eight shale samples from the
Appalachian region was measured using the Jar Slake (Soak) Test. Of
the sixty-eight samples tested, fourteen degraded into a pile of
flakes or mud, four developed small fractures, and fifty experienced
no degradation at all. X-ray diffraction analysis identified
kaolinite as the predominant clay mineral present in the shales.
Shales composed primarily of kaolinite slake as a result of pore air
compression. Pore air compression is favored by small pore radii.
Thin section analysis found the fourteen shales that slaked had a
system of macropores with an average diameter equal to 0.06 mm, the
four samples that developed small fractures had macropores with an
average diameter equal to 0.07 mm, and the fifty samples that
experienced no degradation at all had macropores with an average
diameter equal to 0.092 mm. This indicates that the size of the
macropores has a definite influence on the slaking of the shales.

KEYWORDS: durability, shales, capillary suction, pores, Jar Slake.

----

[1]Associate Professor, Department of Civil Engineering,
University of Pittsburgh, Pittsburgh PA 15261, and Office of Surface
Mining, Ten Parkway Center, Pittsburgh PA 15220
   [2]Geologist, U.S. Office of Surface Mining, 1020 15th Street,
Denver, CO 80202
   [3]Professor of Civil Engineering, Purdue University, West
Lafayette, IN 47907
   [4]Supervisory Physical Scientist, U.S. Office of Surface Mining,
Ten Parkway Center, Pittsburgh, PA 15220

INTRODUCTION

Control of seepage and internal erosion are important design considerations in the construction of embankments. The placement of a durable rock underdrain or system of underdrains is the typical method for minimizing internal drainage problems. Waste fills constructed in the Appalachian region during surface coal mining operations share these same concerns--particularly where shales are employed as the underdrain rock. Thus, the long-term stability of fills can be directly related to the durability of the rock forming the underdrain. If this rock degrades into soil-size particles as a result of moisture absorption, the drainage system provided by the void space between the rocks may become clogged. This clogging will cause the underdrain system to become ineffective and develop excess pore water pressures. These excess pore water pressures cause a decrease in the shear strength of the fill material. This decrease in shear strength can cause the failure of the fill structure. Thus, the correct assessment of the durability of the underdrain rock is a critical factor in coal mining waste fill design when shales are used in the underdrains.

In order to understand the mechanisms involved in the durability of shales, sixty-eight shale samples were collected at recently blasted highwalls at surface mines in Kentucky, Tennessee, Virginia and West Virginia. The samples were subjected in the laboratory to a combination of slake durability tests and petrographic analyses that involved both x-ray diffraction analysis and thin section examination. This study reports the results of the laboratory program and discusses the probable mechanisms influencing the durability of the shales.

DURABILITY TESTS

Several types of tests have been developed or modified to provide qualitative and/or quantitative assessments of the slake potential of geologic materials (Andrews et al. 1980). The durability tests that have been suggested consist of the following: (a) Jar Slake (Soak), (b) Slake Durability, (c) Ultrasonic Degradation, (d) Los Angeles Abrasion, (e) Freeze-Thaw, (f) Rate of Absorption, (g) Swelling, (h) Cyclic Wet-Dry, and (i) Washington Degradation.

In the present study, the Jar Slake (Soak) Test was used to measure the durability of the shales. This test provides a qualitative measure of rock behavior after immersion in water for a 24-hour period. This test is particularly relevant to the durability of underdrain rock in waste fills, as suggested by Andrews et al. (1980) who state, "...this test might be most closely related to spoil materials located at depth and within a constant humidity or totally saturated environment." Strohm et al. (1978) recommend the Jar Slake Test as the "... basic screening test..." for non-durable shales. In addition, this test provides quick and inexpensive results, thus making it very useful for a continuous characterization of the shales (Office of Surface Mining 1992).

Jar Slake data are qualitative, and based on an assessment of the behavior of an as-received or oven-dried sample after a specified period of immersion. A ranking system based on the appearance of the sample after soaking is described by Lutton (1977). The Jar Slake Index (I_j) is ranked on a scale from one to six (Table 1).

TABLE 1   Jar Slake Ranking (Lutton 1977)

Jar Slake Index I_j	Behavior
1	Degrades to a pile of flakes or mud
2	Breaks rapidly and/or forms many chips
3	Breaks slowly and/or forms few chips
4	Breaks rapidly and/or develops several fractures
5	Breaks slowly and/or develops few fractures
6	No change

The Lutton (1977) (Table 1) ranking system was applied to the sixty-eight shale samples forming part of this study. Irregularly-shaped samples weighing at least 100 grams at as-received moisture contents were soaked in distilled water at $20^{\circ}C$ for 24 hours, and then photographed. Pre- and post-test photographs were compared to assess the mode and degree of breakdown. Figures 1 and 2 show the before and after results of the Jar Slake Test conducted on two shale samples, TN-5 and TN-9. The TN-5 sample (Fig. 1) had an I_j value equal to 1 and the TN-9 sample (Fig. 2) had an I_j value equal to 6.

The full range of slaking behavior, from $I_j=1$ to $I_j=6$, was observed in the tested rock. Of the sixty-eight samples tested by the jar slake method, fourteen samples were rated at an I_j of either 1 or 2, four samples were rated at an I_j equal to 3, and fifty samples were rated at an I_j of either 5 or 6. The samples with an I_j of either 1 or 2 disintegrated in the form of many flakes (Fig. 1). Strohm et al. (1978) recommend that "shales with Jar Slake Index, I_j , of 1 or 2 obviously should be considered soil like," with respect to their durability characteristics. Rocks with I_j of 1 or 2 would not meet the regulatory criteria for rock drain material as stated in Federal Regulations for surface coal mining (United States Code of Federal Regulations 1989), which require that rock in these applications do not slake in water and will not degrade to soil material.

In order to understand the mechanisms involved with the disintegration or non-disintegration of the shales when immersed in water, the sixty-eight samples were subjected to petrographic analyses that involved both x-ray diffraction analysis and thin section examination.

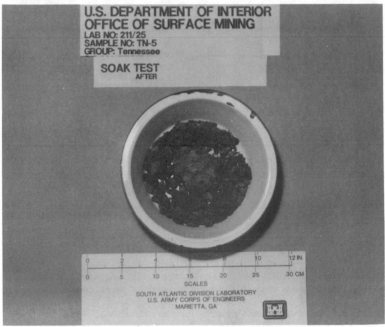

FIG. 1--Before and after Jar Slake Test - Sample TN-5 $(I_j=1)$
(United States Corps of Engineers 1989).

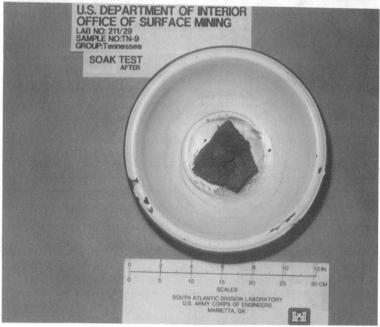

FIG. 2--Before and after Jar Slake Test - Sample TN-9 (I_j=6)
(United States Corps of Engineers 1989).

X-RAY DIFFRACTION AND THIN SECTION ANALYSES

Complete description of shales involves both x-ray diffraction analysis and thin section examination (Muller and Schmincke 1967). X-ray diffraction identifies clay and carbonate minerals, and thin section analysis provides information about the fabric of the shales.

For the x-ray diffraction analysis, the shale samples were ground into powder size material. The material that passed the No. 325 sieve (0.0044 mm size of openings) was x-rayed using a Phillips XRG 3000 Diffractometer. The x-ray scan produces prints called diffractograms. These diffractograms consist of a combination of relatively smooth and peak intensity patterns. The recorded patterns are compared with standard patterns produced by known pure substances in order to obtain the quantities and types of clays and non-clay materials present in the samples. The x-ray diffraction analyses were run on samples that were air dried at room temperature and heat treated at 350 and 575 degrees Celsius. The heat treatment was used because it induces characteristic lattice expansion, contraction and collapse in various clay minerals. These changes are reflected in the peak intensity levels and their location as recorded in the diffractograms of the samples subjected to the heat treatment process. From these changes it is possible to infer the composition of the original material. In addition, to detect the presence of montmorillonite in the samples, x-ray diffraction analysis was undertaken in powder samples exposed to ethylene glycol vapors. Using this method, the presence of expansive clay minerals in the samples is detected by identifying characteristic peak intensity levels and their location in the diffractograms (Muller and Schmincke 1967).

Table 2 shows the results from the x-ray diffraction analysis for the shales from Kentucky, Tennessee, Virginia and West Virginia. An analysis of Table 2 indicates that the most prevalent clay mineral present in the shales was kaolinite. No expansive clay minerals were present in the shales.

As a complement to the x-ray diffraction analysis, a petrographic analysis of the shales was performed. This latter analysis involved the use of thin sections (30 μm in thickness), that were first examined optically with polarizing microscope, and then photographed at two different magnifications (25X and 63X).

SLAKING MECHANISMS

There are different mechanisms discussed in the geotechnical literature which explain the slaking of shales when immersed in water. One slaking phenomenon is attributed to the compression of entrapped air in the pores of the shale when water enters the shales as a result of capillary suction (Moriwaki 1974). This entrapped air in the pores exerts tension on the solid skeleton, causing the material to fail in tension. According to Moriwaki (1974), pore air compression is the

TABLE 2   X-Ray Diffraction Analysis

Shales Origin (# of samples)	Kaolinite (%)	Quartz (%)	Mica (%)	Feldespar (%)	Other* (%)
Kentucky (30)	18	26	9	17	30
Tennessee (12)	6	42	26	5	21
Virginia (11)	8	30	37	6	19
West Virginia (15)	23	26	3	25	23

* Other includes: Chlorite, Ankerite, Calcite, Siderites and Opaques.

predominant slaking mechanism in shales composed primarily of non-expansive clay minerals such as kaolinite. Clay surface hydration by ion adsorption has been suggested as another mechanism that causes slaking through the swelling of montmorillonite clays in the shales (Andrews et al. 1980). The removal of cementing agents from the shales by groundwater dissolution is also considered to be a mechanism that causes slaking (Surendra et al. 1981).

Since the sixty-eight shale samples tested in the Jar Slake (Soak) Test have kaolinite as the primary clay mineral in the structure, pore air compression should be (according to Moriwaki 1974) the primary mechanism causing the slaking of the shales. The Jar Slake Tests resulted in fourteen samples completely disintegrating when immersed in water; four samples developing small fractures; and fifty samples experiencing little or no change after soaking in water. In order to understand the reasons for the slaking and non-slaking behavior of the shales, an analysis of their fabric was undertaken using the photographs of the thin sections.

PORE STRUCTURE

Pore Pressures

Shales with non-expansive clays (i.e. kaolinite clay) slake because of pore air compression (Moriwaki 1974). Pore air compression is exhibited when water is drawn into the macropore system of the shale during the immersion process of the Jar Slake Test. The suction of water by the shales is the result of capillary forces. This suction process is illustrated by Fig. 3, which represents a shale sample with a system of macropores that run continuously through the sample (Fig. 3(A)). These macropores, which are assumed not to interconnect, resemble small tubes inside the shale. When the sample is immersed in water, water will be pulled into the individual

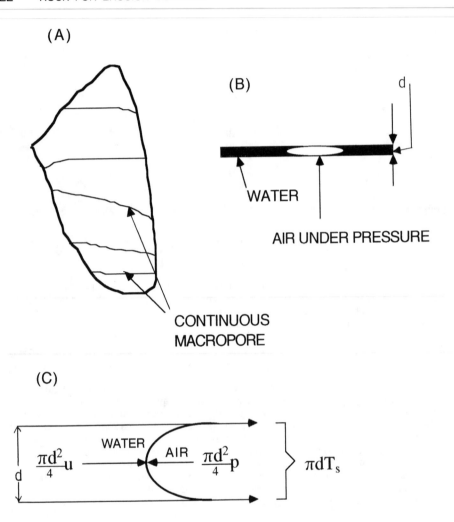

FIG. 3 -- (A) Shale sample; (B) Macropore with water
           and air pressures; (C) Air and water forces
           at the air-water interface in a macropore
           (Juarez-Badillo and Rico-Rodriguez 1986).

macropore, as a result of capillary forces, and the air that
originally filled the macropores will be subjected to compression
(Fig. 3(B)). The system of forces acting at the interface between the
air and the water in the macropore are modeled in Fig. 3(C) (Juarez-
Badillo and Rico-Rodriguez 1986). According to Juarez-Badillo and
Rico-Rodriguez (1986), the following equation applies at equilibrium
conditions:

$$\pi d T_s - \frac{\pi d^2}{4} p + \frac{\pi d^2}{4} u = 0$$

(1)

where, d = the diameter of the macropore
           T_s = the surface tension of water acting on the meniscus
           p = the air pressure
           u = pore water pressure

From Equation (1) the following relationship can be obtained

$$p = u + \frac{4T_s}{d}$$

(2)

An analysis of Eq. (2) indicates that the pore pressure, p , in
the portion of the macropore filled with air (Fig. 3(B)) increases as
the diameter, d , of the macropore decreases. Thus, the smaller the
diameter of the macropore, the larger the air pressure will be.
Since pore air compression is favored by small pore radii, slaking of
shales by air compression will be more pronounced in those shales
containing small diameter macropores. In addition, small diameter
macropores will more readily confine the air pressure developed during
the suction process. That is, diffusion of the air pressure will
decrease with the decrease in surface area (which is a function of the
diameter of the macropore) of the pore that is in contact with the
air.

Moriwaki (1974) also stressed the point that air compression as
a slaking mechanism will be more effective in materials which have a
low degree of water saturation before they are immersed in water.
Moriwaki (1974) determined that the lower the degree of saturation of
shales, the more rapidly was water drawn into the pores. This rapid
absorption of water by the shales will prevent the air pressures from
dissipating, hence making them more effective in breaking the shales.
The sixty-eight samples in this study had very low water contents (as
received water content in Table 3) before they were soaked in the Jar
Slake Test. Thus, slaking behavior may have been attenuated by the
small amount of water in the macropores. Therefore, for optimum
results when using the Jar Slake Test, it is recommended that the
samples be oven-dried before immersion in water (Office of Surface

Mining 1992).

Table 3. Average Water Content in Shale Samples

Origin of Samples	Number	As Received Water Content %	Water Content, Saturated Conditions %
Kentucky	30	0.5	4.2
Tennessee	12	1.2	3.2
Virginia	11	0.8	2.6
West Virginia	15	0.7	3.8

From the previous discussion it can be concluded that the diameter, and to a lesser extent, the degree of water saturation of the macropore system of the shales has a marked influence on their slaking in water.

Pore Diameter

In order to study the pore geometry of the sixty-eight shale samples, thin sections (30 μm in thickness) were prepared and photographs of the sections were made using a polarizing microscope. The study of pore space geometry of rocks using photographs of thin sections is a standard procedure in petrographic analysis (Blatt et al. 1972; Dullien 1979; Schlueter et al. 1991). The photographs of the thin sections were made at two different magnifications (25X and 63X).

From the photographs of the thin sections, the cross-sectional areas and the shape of the perimeters of the macropores in the shales were obtained. To obtain the areas of the macropores as well as a plot of their perimeters, the macropores were determined by standard digitizing procedures (Starkey and Simigian 1987; Stanton 1987). Digitizing records the x-y coordinates of the perimeter of each macropore. This information is analyzed by a microcomputer with software necessary to calculate the areas of the macropores and to plot their perimeter. Figure 4 shows the typical results of the process for shale samples TN-5 and TN-9.

Using the area of each macropore, the "equivalent diameter" of the macropore was obtained. This "equivalent diameter" corresponds to the diameter of a circular area that has the same area as the macropore. From each photograph, forty representative macropores were chosen to calculate the overall average "equivalent diameter" for each shale sample. For example, the average "equivalent diameter" for the forty macropores of sample TN-5 (Fig. 4(A)) is equal to 0.053 mm. In the case of sample TN-9 (Fig. 4(B)), this overall average "equivalent

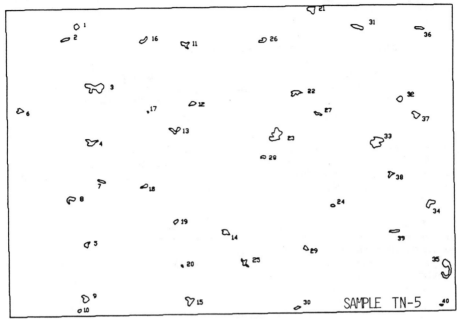

scale: _____ 1 mm _____

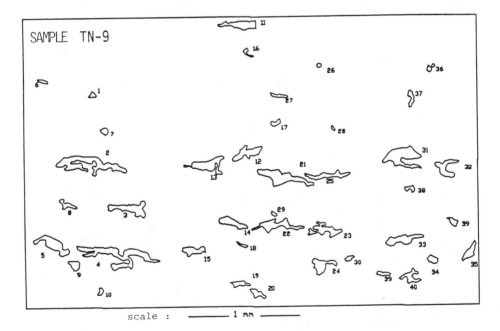

scale : _____ 1 mm _____

FIG. 4--Pore Geometry in samples TN-5 and TN-9.

diameter" is equal to 0.107 mm.

RELATIONSHIP BETWEEN PORE DIAMETER AND DEGREE OF SLAKING

As previously mentioned, from the Jar Slake Test on the sixty-eight shale samples, fourteen of them completely disintegrated during this test and had a Jar Slake Index (I_j) of 1 or 2 (Table 1); four samples developed small fractures and had an I_j equal to 3; and fifty shale samples experienced no degradation at all and had an I_j value of either 5 or 6. Thus, three groups of shales were established with respect to their durability. One group of fourteen samples disintegrated into soil-size particles, a second group of four samples experienced very small changes, and a third group of fifty samples suffered no change when immersed in water.

Next, the mean macropore diameters of each of these three groups were calculated and compared with their slaking behavior. This was done in order to determine if there is a correlation between macropore diameter and their degree of slaking in water. The results of these calculations are shown in Table 4.

Table 4. Slaking Behavior and Pore Diameter

Number Of Samples	Slake Index I_j	Maximum Diameter mm	Minimum Diameter mm	Standard Deviation	Mean Diameter mm
14	1 or 2	0.081	0.043	0.0125	0.060
4	3	0.082	0.050	0.0136	0.070
50	5 or 6	0.180	0.050	0.030	0.092

An analysis of Table 4 indicates that the fourteen samples that completely disintegrated during the Jar Slake Test had a system of macropores with an average diameter equal to 0.06 mm, the four samples that developed small fractures had macropores with an average diameter equal to 0.07 mm, and the fifty samples that experienced no degradation al all had macropores with an average diameter equal to 0.092 mm. Thus, there appears to be an optimum macropore size at which air pressures are effective in breaking the shales.

CONCLUSIONS

The durability of sixty-eight shale samples from the Appalachian region was analyzed using a combination of slake

durability tests, petrographic analyses that included x-ray
diffraction and thin section examination, and computer assisted image
analysis of the macropore system of the shales. From this combination
of tests, the following determinations were made:

1. The shales had the non-expansive clay, kaolinite, as the
predominant clay mineral in their structure.

2. The predominant slaking mechanism of the shales was pore
air compression that takes place when the shales are immersed in water
and is the result of capillary suction pressures.

3. Slaking by pore air compression was directly related to the
average diameter of the macropore system in the shales. The smaller
the diameter, the more pronounced was the slaking of the shales by
pore air compression.

4. There appears to be an optimum macropore diameter at which
air pressures are most effective in breaking the shales. For the
shales tested this average diameter seems to be equal or smaller than
0.06 mm.

ACKNOWLEDGEMENTS

The authors give special thanks to the U.S. Army Corps of
Engineers Missouri River and South Atlantic Divisions in whose
laboratories the testing program described in this study was carried
out. Special thanks are also given to Ann Stewart-Murphy of the
Office of Surface Mining, Pittsburgh, for conducting the computer
analysis of the pore geometry of the shales.

REFERENCES

Andrews, D.E., Withiam, J.L., Perry, E.F., and Crouse, H.L., 1980,
    Environmental Effects of Slaking of Surface Mine Spoils:
    Eastern and Central United States, Final Report, U.S. Bureau
    of Mines, U.S. Department of the Interior, Denver, CO., 247 p.

Blatt, H., Middleton, G., and Murray, R., 1972, Origin of Sedimentary
    Rocks, Prentice-Hall Inc., Englewood Cliffs, New Jersey.

Dullien, F.A.L., 1979, Porous Media:Fluid Transport and Pore
    Structure, Academic Press, New York.

Juarez-Badillo, E., and Rico-Rodriguez, A., 1986, Soil Mechanics,
    Vol. I, Limusa Publishing Co., Mexico.

Lutton, R.J., 1977, Design and Construction of Compacted Shale
    Embankments: Slaking Indices for Design, Report FHWA-RD-77-1,
    Federal Highways Administration, Washington, D.C., 88 p.

Moriwaki, Y., 1974, Causes of Slaking in Argillaceous Materials, Ph.D.

Dissertation, Department of Civil Engineering, University of California at Berkeley, Berkeley, CA., 291 p.

Muller, G., and Schmincke, H., 1967, *Methods in Sedimentary Petrology*, Hafner Publishing Co., New York.

Office of Surface Mining, 1992, *Overburden Strength-Durability Classification System for Surface Coal Mining*," Open File Report, Office of Surface Mining, U.S. Department of the Interior, Pittsburgh, PA., 45 p.

Schlueter, E.M., Cook, N.G.W., Zimmerman, R.W., and Witherspoon, P.A., 1991, "Predicting Permeability and Electrical Conductivity of Sedimentary Rocks from Microgeometry," *Proceedings*, 32nd U.S. Symposium on Rock Mechanics, Norman, Oklahoma, pp.355-364.

Stanton, T., 1987, "Tablets for Precision Graphics," *PC Magazine*, Vol.6, pp. 159-181.

Starkey, J., and Simigian, S., 1987, "Image: A Fortran V Program for Image Analysis of Particles," *Computers and Geoscience*, Vol. 13, pp. 37-59.

Strohm, W.E., Bragg, G.H., and Ziegler, T.W., 1978, *Design and Construction of Compacted Shale Embankments*, Report HWA-RE-78-141, Federal Highway Administration, Washington, D.C., 207 p.

Surendra, M., Lovell, C.W., and Wood, L.E., 1981, "Laboratory Studies of the Stabilization of Nondurable Shales," *Transportation Research Record*, Vol. 790, pp. 33-40.

United States Code of Federal Regulations, 1989, *Title 30, Mineral Resources, Chapter VII, Enforcement*, Department of the Interior, Sections 817.71 and 817.73, pp. 320-324.

United States Corps of Engineers, 1989, *Mechanical and Petrographic Analysis of Rockfill Materials*, South Atlantic Division Laboratory Report, Marietta, Georgia, 175 p.

Terry L. Johnson,[1] Steven R. Abt,[2] and Myron H. Fliegel[3]

VERIFICATION OF ROCK DURABILITY EVALUATION PROCEDURES USING PETROGLYPHS
AND INDIAN ROCK ART

REFERENCE: Johnson, T. L., Abt, S. R., and Fliegel, M. H., **"Verification of Rock Durability Evaluation Procedures using Petroglyphs
and Indian Rock Art,"** Rock for Erosion Control, ASTM STP 1177,
Charles H. McElroy and David A. Lienhart, Eds., American Society for
Testing and Materials, Philadelphia, 1993.

ABSTRACT: Rock riprap erosion protection is normally used at
radioactive waste disposal sites to stabilize waste covers against wind
and water erosion. Federal statutes require that the covers and erosion
protection remain effective for hundreds of years, without placing
reliance on active maintenance. To provide guidance in selecting rock
sources to provide this level of protection, the Nuclear Regulatory
Commission (NRC) staff developed a quantitative procedure for evaluating
the physical characteristics of rock. To refine and improve this
procedure, a contract was awarded to Colorado State University (CSU) to
gather data on rock types which have survived for long periods of time.
Emphasis was placed on obtaining data from marginal-quality sandstone
rocks found at petroglyph and rock art sites in the southwestern United
States. Rock samples near the petroglyphs were gathered, and standard
durability tests were performed on the samples; archaeological
information was used to determine the age of the petroglyphs. Even
though the information gathered may only be applicable to arid and semi-
arid areas, previously-developed NRC testing and scoring procedures will
likely be modified and refined as aditional data become available.

KEYWORDS: Rock durability, riprap testing

INTRODUCTION

Low-level waste and uranium mill tailings waste disposal sites
normally employ protective cover systems to limit radioactive releases
and to prevent erosion and dispersion of contaminated material. These
cover systems are generally composed of a clay cover protected by a
another layer which minimizes the potential for gullying and erosion
damage.

---

[1]Senior Hydraulic Engineer, United States Nuclear Regulatory Commission,
Washington, D.C., 20555.

[2]Professor, Civil Engineering Department, Colorado State University, Fort
Collins, CO, 80523.

[3]Section Leader, United States Nuclear Regulatory Commission, Washington,
D.C., 20555.

Various regulations establish several major design objectives for long-term stabilization at radioactive waste disposal sites. These objectives can be summarized as follows: (1) prevent transport of contaminated material caused by wind and water erosion; (2) provide long-term stability without placing reliance on active maintenance; and (3) provide sufficient protection to limit radioactive releases. Because vegetation alone is often not effective in many areas, and because natural steep slopes are common on small watersheds, rock riprap is often needed to provide adequate erosion protection. At a typical waste disposal site, riprap may be needed to protect top and side slopes, diversion channels, aprons and diversion channel outlets, and banks of large rivers or areas where floods impinge. Federal regulations, such as 10 CFR Part 40 and 10 CFR Part 61, require that the cover system be effective for time periods ranging from 200 to 1000 years, or more.

Of particular importance in the design of riprap covers is the need to provide long-term stability without placing reliance on active maintenance. To meet this requirement, the rock riprap layer must not be susceptible to significant weathering and must remain functional over a long period of time. Therefore, it can be seen that rock covers must survive natural weathering forces for hundreds of years.

The historical basis for evaluating rock durability is well-developed. Jahns (1982) points out that many kinds of rocks are relatively resistant to weathering. Most of these more resistant rock types have long been used as construction materials, in monuments, or for decorative purposes, with varying degrees of success. However, most engineers and designers recognize that there are limitations associated with procedures that are used to assess rock performance over long time periods.

Determining the quality of riprap needed for long-term protection and stability can therefore be a somewhat difficult and subjective task. Very little design guidance is available to assess rock quality, based on physical properties. Considerable engineering judgment is necessary to develop rational engineering design alternatives when weathering of rock is a major consideration. Any rational design method must include the durability and weathering characteristics of the rock in the analysis.

In assessing the long-term durability of erosion protection, the NRC staff has relied on the results of durability tests performed at several waste disposal sites and on information and analyses developed by technical assistance contractors. Methods and procedures were developed in these studies which considered field data and provided criteria for meeting long-term stability requirements, based on the results of physical durability tests.

The NRC staff recently developed a quantitative procedure to assess the quality of rock used for erosion protection at waste disposal sites (NRC 1990). Development of the procedure was based on experience with various reclamation designs and on published rock quality procedures. Recognizing that little data are currently available in this area, the NRC staff awarded a contract to Colorado State University (CSU) to obtain additional information that will be used to refine and improve the procedure.

DISCUSSION

The ability of some rock types to survive exposure and weathering without significant degradation for long time periods is well-documented by archaeological and historic evidence (Lindsey et al. 1982; Jahns 1982). However, little information is available to quantitatively assess the quality of rock needed to survive for long periods of time, based on its physical properties. In previous assessments, the NRC staff relied on information and analyses suggested by Nelson et al. (1986). Refinements were made to these methods and were published in NRC Final Staff Technical Position (FSTP) "Design of Erosion Protection Covers for Stabilization of Uranium Mill Tailiings Sites" (NRC 1990).

The NRC staff developed a scoring procedure to quantitatively evaluate rock proposed for use as riprap. The procedure involves using a suite of durability tests, recognizing that no single test can always be relied on to determine rock quality. The result of a durability test is correlated to a rock quality score and then multiplied by a weighting factor, depending on the type of rock (See Table 1). The final rock quality rating is derived by summing the weighted scores and determining the percentage of the maximum possible score for that type of rock.

To show how the scoring procedure is applied, a hypothetical example for a sandstone is presented below. The example provides information on the conversion of physical test data to scores, application of weighting factors, and derivation of an overall rock quality rating.

Test		Result	Score	Weight	Score x wgt.	Max
Spec. Gravity	(ASTM C127)	2.61	7	6	42	60
Absorption	(ASTM C127)	1.22 %	4	5	20	50
Sod. Sulfate	(ASTM C88)	6.90 %	6	3	18	30
L.A. Abrasion	(ASTM C131)	8.70 %	5	8	40	80
Schmidt Hammer		51	6	13	78	130
Tensile Str.	(ASTM D3967)	670 psi	6	4	24	40
Totals					222	390

The overall rock quality rating for this hypothetical sandstone would be computed as 222/390 or 57 percent. Such rock would be considered to be of "fair" quality.

The numerical scores in Table 1, corresponding to a particular test result, were derived from published data (Lindsey, et al. 1982). The scores represent the relative rock quality on a scale of 0 to 10. Classifications of poor (0-5), fair (5-8), and good (8-10) were given actual numerical values, based on that data.

The weighting factors presented in Table 1 were derived from Table 7 of "Petrographic Investigations of Rock Durability and Comparisons of Various Test Procedures," (Dupuy 1965). In that report, Dupuy ranked 13 different standard durability tests according to their applicability for limestones, sandstones, and igneous rocks. The weighting factors in Table 1 are based on an inverse ordering of Dupuy's ranking of test methods for each rock type. For example, the test considered to be most applicable for a specific type of rock is assigned a weighting factor of 13 and the least applicable test receives a weighting factor of 1. Tests other than those listed in Table 1 may be used, and the weighting factors may be derived by counting upward from the bottom of Table 7 in Dupuy's report.

The acceptability of the rock is also dependent on the results of a petrographic examination. To be acceptable for long-term use, the rock must be qualitatively rated at least "fair" and must not contain substantial amounts of smectites and expanding clay-lattice minerals.

TABLE 1

Scoring Criteria for Determining Rock Quality

Laboratory Test	Weighting Factor			Score										
	Limestone	Sandstone	Igneous	10 (Good)	9 (Good)	8	7	6 (Fair)	5 (Fair)	4	3 (Poor)	2 (Poor)	1	0
Sp. Gravity	12	6	9	2.75	2.70	2.65	2.60	2.55	2.50	2.45	2.40	2.35	2.30	2.25
Absorption, %	13	5	2	.1	.3	.5	.67	.83	1.0	1.5	2.0	2.5	3.0	3.5
Sodium Sulfate, %	4	3	11	1.0	3.0	5.0	6.7	8.3	10.0	12.5	15.0	20.0	25.0	30.0
L/A Abrasion (100 revs), %	1	8	1	1.0	3.0	5.0	6.7	8.3	10.0	12.5	15.0	20.0	25.0	30.0
Schmidt Hammer	11	13	3	70.0	65.0	60.0	54.0	47.0	40.0	32.0	24.0	16.0	8.0	0.0
Tensile Strength, psi	6	4	10	1400	1200	1000	833	666	500	400	300	200	100	0

Esmiol (1967) found that one of the principal causes of riprap failure was the presence of clay minerals, which tend to break down rocks when exposed to weathering cycles or natural stresses. Following the petrographic examination, the acceptability of the rock is dependent on the overall rating and where it will be used in the design. Rock that is to be used in a critical area requires a higher rating than rock used in other areas. The NRC procedure defines minimum ratings needed in particular areas. In some situations, rock that fails to meet that suggested criteria can still be used, if it is properly oversized to compensate for its expected degradation.

Oversizing criteria vary, depending on the location where the rock will be placed. Areas that are frequently saturated are generally more vulnerable to weathering than occasionally-saturated areas where freeze/thaw and wet/dry cycles occur less frequently. For rock oversizing purposes, the following criteria have been developed:

A. Critical Areas.   These areas include, as a minimum,
                          frequently-saturated areas, channels,
                          toes, aprons, etc.

Rating	Oversizing Criteria
80-100	No oversizing needed
65-80	Oversize using factor of (80 - Rating)
Less than 65	Reject

B. Non-Critical Areas.   These areas include occasionally-
                          saturated areas, top slopes, side
                          slopes, and well-drained areas.

Rating	Oversizing Criteria
80-100	No oversizing needed
50-80	Oversize using factor of (80 - Rating)
Less than 50	Reject

At many sites, it may be difficult to locate nearby sources of high-quality rock. The NRC staff has reviewed several reclamation designs where marginal-quality sandstones were proposed for use, due to the difficulty in locating good-quality sources. When the scoring procedure was used to evaluate such sandstones, the scores did not meet the minimum criteria. Use of alternate sources located a substantial distance from the site resulted in designs which were impractical to construct.

Staff experience with reclamation and closure designs has indicated that this procedure is capable of easily distinguishing rock of very good or very poor quality. This conclusion is based on qualitative examination of rocks that appear to be of relatively good and relatively poor quality. It is fairly easy to conclude that certain types of sandstones (e.g., Navajo, Bluff, and Wingate) are of generally poor quality and are easily weathered when exposed. Also, it is easy to conclude that certain types of basalts and quartzites will not weather significantly over a long period of time. However, rock of marginal quality (scoring approximately 40 to 50 using the NRC procedure) is often questionable. Staff experience indicates that the procedure generally produces conservative results when applied to sandstones. While conservatism is appropriate when designing for natural events over a long period of time, there is also a need to provide criteria that considers any weathering data that are applicable to long-term applications. In general, quantitative data do not exist in this area, and the staff recognizes that additional information is needed to verify and validate the scoring and oversizing procedures, particularly for marginal-quality rock.

ANALYSIS

To provide additional verification of the NRC procedure, a contract was awarded to CSU to gather additional data. It was determined that this additional data should be collected to verify, if possible, weathering rates and long-term performance of various types of rocks, based on their physical properties. It was reasoned that rocks could be found where little or no weathering occurred and where the age of the rock could be determined; the most difficult part of the project would be to locate rocks with a known exposure period. This was accomplished by locating Indian petroglyphs and rock art, where the age had been previously determined by archaeological studies. Sites were considered that contained petroglyphs pecked into desert varnish; thus, the minimum time of exposure could be estimated. (Formation of a rock varnish is a slow process in which successive layers are deposited over a very long period of time. During wet periods, bacteria flourish on the surface of the rock, producing oxides of manganese and iron. The metal oxides cement clay minerals into a dark varnish on the surface of the rock.) It was assumed that nearby similar rocks in the area with similar desert varnish, but no petroglyphs, would be at least that old. Since the rocks had survived relatively intact and still had the desert varnish, it would be reasonable to measure the physical properties of the nearby rocks to determine those factors which contribute to long-term stability.

Physical properties of rocks at various sites were measured using standard durability tests and were assigned scores, based on the current NRC criteria. Samples were taken from areas where the archaeological features were openly exposed to weathering and from areas where direct exposure did not occur (i.e. the petroglyphs were located in sheltered areas such as rock overhangs and/or alcoves). Following is a list of sites that have been evaluated:

Sample Location	Rock Type
Exposed	
Chaco Canyon, NM	Pictured Cliffs Sandstone
Escalante Canyon, CO	Dakota Sandstone
Shavano Valley, CO	Dakota/Burro Canyon Sandstone
Cedar Fort, UT	Weber(?) Sandstone
Sheltered	
Sand Island, UT	Navajo Sandstone
Sego Canyon, UT	Navajo Sandstone

The two sheltered rock sites also contained pictographs, in

addition to petroglyphs. In these locations, Indian rock art, in the form of paintings, had survived for many hundreds of years. It became obvious that samples taken from these areas would not represent the same degree of exposure to weathering that would simulate weathering of rock riprap. The rock had not been exposed to sufficient weathering to remove even paint from the rock. A decision was made to limit the testing of these types of samples, since the results would not be very meaningful. Nevertheless, to complete the data base, it was concluded that it would be interesting to determine if even poor-quality rock could survive if it was somewhat sheltered. To accomplish this, a sample from the Sego Canyon site was tested. The final rock quality score was determined to be 17 out of a possible 100, indicating that poor-quality rock could survive for long periods if sheltered from exposure.

A summary of test results for various samples of exposed rocks is presented in Table 2.

TABLE 2
DURABILITY TEST RESULTS
EXPOSED SANDSTONE SAMPLES

Test (weight)	Chaco	Escalante	Shavano	Cedar Fort
Specific Gravity (6)				
Test Result	2.67	2.52	2.50	2.48
Score	8.4	5.4	5.0	4.6
Absorption (5)				
Test Result	1.6	2.3	2.3	2.5
Score	9.0	2.4	2.4	2.0
NaSO4 (3)				
Test Result	30	4.2	3.2	4.4
Score	0	8.4	8.9	8.3
L. A. Abrasion (8)				
Test Result	11.0	7.6	10.6	15.4
Score	4.0	6.4	4.8	3.0
Schmidt Hammer (13)				
Test Result	27	25	24	32
Score	3	3	3	4
Tensile Strength (4)				
Test Result	990	1450	1900	1600
Score	7	10	10	10
Totals				
Total Weighted Scores	194	200	186	178
Maximum Score	390	390	390	390
Rock Quality Rating	50	51	48	46

From a qualitative standpoint, the authors are relatively confident that the samples collected at Escalante Canyon, Shavano Valley, and Cedar Fort are relatively durable sandstones. The scores of 51, 48, and 46 intuitively seem to be too low, because the rocks have survived prolonged exposure and weathering for approximately 600-1000 years (Castleton 1987) without noticeable changes, based on examination of the nearby petroglyphs. Thus, it appears that some refinements may be needed to the NRC scoring procedure to reflect the fact that certain sandstones are able to survive long periods of exposure. While several different options are available for revising the criteria, the task is more complicated than simply lowering the minimum acceptable score of 50 to a minimum of approximately 40 or 45. The authors are aware of rock testing data where rocks that were considered to be of very poor quality scored 45-55. Thus, this information also needs to be considered in the evaluation.

One possible approach is to modify the weighting factors, based on the preliminary data. It appears that the Schmidt Hammer test is given too much weight (13) and the Sodium Sulfate Soundness test is given too little weight (3), since the high-confidence rock samples scored well in the Sulfate Soundness test and poorly in the Schmidt Hammer test. Therefore, consideration will be given to changing the weighting factors for these tests. There are many different changes that could be made. For example, if the Schmidt Hammer test is arbitrarily assigned the minimum weight of 1 and the Sodium Sulfate test is arbitrarily assigned the maximum weight of 13, the scores for the Escalante, Shavano, and Cedar Fort samples would be changed to 67, 65, and 58, respectively. Such scores would make the rock suitable for use in long-term applications. The scores for the poorer-quality rock samples remain relatively unchanged.

Based on the preliminary data, it appears that some changes may be made to the NRC scoring procedure for sandstone. However, additional research will be conducted and additional test data will be collected. Future plans are to investigate additional petroglyphs and Indian ruins to expand the rock scoring data base and to possibly verify the oversizing criteria by quantitatively measuring physical deterioration of Indian ruins. This will be accomplished by measuring, for example, the sizes of sandstone blocks used to construct habitation sites that are sheltered and comparing the results to nearby similarly-constructed structures that are openly exposed to weathering. In addition, the applicability of the revised procedure will be assessed for areas other than the southwestern United States.

CONCLUSIONS

Preliminary results of durability tests of marginal-quality sandstones at Indian petroglyph sites indicate that the NRC scoring criteria developed in the FSTP may need some refinement for use with marginal-quality sandstones. The project for collection of data to refine the criteria is in its early stages; however, it appears that several durability tests may not be good indicators of rock durability for marginal-quality sandstones. The NRC staff intends to continue research to collect data at Indian ruin sites, expand the data base, and to refine the scoring and oversizing procedures currently used for long-term erosion protection design.

REFERENCES

Castleton, Kenneth B., _Petroglyphs and Pictographs of Utah_, Volume 2, Utah Museum of Natural History, Salt Lake City, 1987.

Dupuy, G. W., "Petrographic Investigations of Rock Durability and Comparisons of Various Test Procedures," _Engineering Geology_, July, 1965.

Esmiol, Elbert E., "Rock as Upstream Slope Protection for Earth Dams - 149 Case Histories," U. S. Bureau of Reclamation, Report No. DD-3,

September, 1967.

Jahns, R. H., "Weathering and Long-Term Durability of Rock Masses," in NUREG/CR-2642, U. S. Nuclear Regulatory Commission, Washington, D.C., 1982.

Lindsey, C. G., L. W. Long, and C. W. Gegej, "Long-Term Survivability of Riprap for Armoring Uranium Mill Tailings and Covers: A Literature Review," NUREG/CR-2642 (PNL-4225), U. S. Nuclear Regulatory Commission, Washington, D.C., 1982.

Nelson, et al., "Methodologies for Evaluating Long-Term Stabilization Designs of Uranium Mill Tailings Impoundments," NUREG/CR-4620, 1986.

U. S. Nuclear Regulatory Commission, Final Staff Technical Position (FSTP), "Design of Erosion Protection Covers for Stabilization of Uranium Mill Tailings Sites," 1990.

Srinivas S. Yerrapragada,[1] Sanjeev S. Tambe,[1] and K.L. Gauri[2]

FRACTALS, PORE POTENTIAL, AND SPHINX LIMESTONE DURABILITY

REFERENCE: Yerrapragada, S. S., Tambe, S. S., and Gauri, K. L., "Fractals, Pore Potential, and Sphinx Limestone Durability," Rock for Erosion Control, ASTM STP 1177, Charles H. McElroy and David A. Lienhart, Eds., American Society for Testing and Materials, Philadelphia, 1993.

ABSTRACT: A study of the porosimetric data of limestone samples from the region of the Sphinx reveals that the pore volume distributions obey a power law. It should be possible, therefore, to treat these strata as a fractal structure. We found that each lithologic unit of the Sphinx, below the neck, can be characterized by a fractal dimension D, the values of which range from 2.54 to 2.77. The volume of small pores, radii $< 0.9\mu m$, ranges from 19% for D = 2.54 to 61% for D = 2.77.
A good correlation, with a correlation coefficient of 0.97, was found between the durability factors as previously determined through pore-size distributions and those determined through the fractal dimension.

KEYWORDS: durability, limestone, fractal dimension, pore-size distributions, pore potential

INTRODUCTION

The durability of stone is conventionally determined through natural exposure tests (Kessler and Anderson 1951) or accelerated weathering (ASTM Standard C-666 1973). However, such tests are time consuming and frequently do not reflect the true response of the material to prolonged outdoor weathering. This paper presents a novel method to characterize the durability of the Sphinx limestone by using fractal dimension.

The limestone of the Sphinx is a biomicrite. The lower strata are sparse-foramininiferal marly layers and grade upwards into a packed biomicrite. Further, lower portion of each stratum noted i in Fig. 1, is richer in clay and extremely fine quartz dust than the upper portion, noted ii. These limestone strata are differentially weathered as seen from the profile where protruded and recessed layers alternate showing respectively higher and lesser degree of weathering (Fig. 1). Gauri (1984) correlated the higher degree of weathering with the lithology of

[1]Research Associate, Geology Department, University of Louisville, Louisville, KY 40292.

[2]Professor, Geology Department, University of Louisville, Louisville, KY 40292.

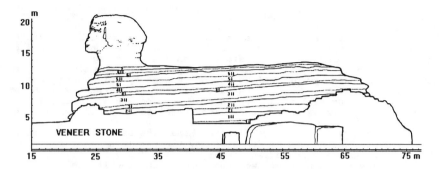

FIG. 1--A photogrammetric view of the Sphinx from North showing
the beds, 1-7, and their relative weathering.

the strata and greater proportion of halite and gypsum present in more
weathered rock. He argued that the salts were of synsedimentary origin
and were thus initially uniformly, though sparsely, distributed
throughout each stratum. These salts cause disintegration due to the
crystallization pressures they generate in the surface region where they
have become concentrated due to the evaporation of water.

Further studies on weathering of the Sphinx limestone revealed
that the stress generated in the pores due to salt crystallization
depends upon the pore-size distributions (Gauri and Punuru 1989). Stones
with large microporosity were found to be more vulnerable than those
with large macroporosity and the pressure change (dp) can be expressed
as

$$dp = 2\Gamma((\frac{dA}{dV})_r - (\frac{dA}{dV})_R)$$ (1)

where
  Γ = surface tension of the solution, N/m,
  r = radius of interconnected small pores, m,
  dA = area of pores in a given size range, m^2, and
  dV = volume of pores in a given size range, m^3.

Applying this equation, it was found that in the lower beds with
large microporosity one crystallization cycle could produce as much as 4
MPa (600 psia) pressure as compared with nearly 1.4 MPa (200 psia)
produced in the upper strata. Following this, Gauri and Punuru (1989)
and Punuru et al. (1990) introduced an expression for the durability
factor of different limestone beds of the Great Sphinx as

$$DF = A_1 V_1 + A_2 V_2 + A_3 V_3$$ (2)

where
                 DF = durability factor for a given stratum,
  V_1, V_2, V_3 = percentage pore volumes in the pore ranges of >5,
                 0.5-5, <5 μm respectively, and
  A_1, A_2, A_3 = constants with values 1.2338, 2.622, and -0.9841.

They obtained values for these coefficients by assigning empirically
high values of 96, 98, and 100 to increasingly more sound, three well

preserved strata as observed at the Sphinx. The above expression was
found to apply to other strata of the Sphinx.
        The Washburn equation used to calculate the pore sizes given above
assumes that the pores are cylindrical. Further, the contact angle to
calculate the pore-size was considered constant both for the entry and
retreat of the mercury into and from the pores. The anomaly likely to be
produced by these approximations can be reduced by calculating the
thermodynamic work, PV, which can also be used to characterize the stone
and calculate durability factors through pore potential (Gauri and
Yerrapragada in press). These DF also correlate very well with those
given by Eq 2.
        In the present study we find that the pore-size distribution of
the Sphinx limestone can be described by the unitless fractal dimension.
Unlike the customary Eucledian linear, planar, and volumetric objects
which are described by one, two, and three axes, the fractal dimension
is described by a number between 1 and 2 for curved lines and two and
three for objects that are neither conventionally planar nor
geometrically three dimensional, such as pores in a rock. The fractal
dimension obeys a power law and can be defined as N α $(1/r^D)$, where N is
the number of objects associated with the size r and D is the fractal
dimension (Turcotte 1991). In the fractal analysis given below, the
pattern of the calculated fractal dimensions follow strictly the pattern
of durability factors calculated through pore-size distributions and the
PV-work.
        The present study pertains mainly to the strata of the Sphinx, but
was found to apply to certain other limestone such as the Indiana
Limestone. The applicability of the DF to other limestone should be
checked.

METHODS

        The details of the samples and the application of the mercury
porosimetry to obtain pore-volume data are given in Punuru et al.
(1990). That data has been utilized in our present analysis.
        Several pressure-volume points were obtained by the porosimeter
measurements. The pore-volumes for the pressure points between the
experimental points were generated by spline interpolation algorithm.
Thus continuous pore volume distribution plots were constructed. The
portion of these distribution curves which exhibited fractal behaviour
were subsequently treated by linear regression analysis to obtain
fractal dimensions.

FRACTAL MODEL AND THE SPHINX LIMESTONE

        The following expression is adapted from Friesen and Mikula (1987)
relating the pore volumes (v), pore radii (r) and fractal dimension (D):

$$\log(-\frac{dv}{dr}) \quad \alpha \quad (2-D)\log(r) \qquad (3)$$

        In applying this equation to our case, we determined the pore
volume and pore radii of the various strata of the Great Sphinx through
mercury intrusion porosimetry measurements. We used the classical
Washburn equation to determine pore radii and applied pressure (P) as

$$r = \frac{2\Gamma|\cos\theta|}{P} \qquad (4)$$

where
        θ = contact angle of mercury with the stone surface.

Then, combining the expressions (Eqs 3 and 4) we get the result that

$$\log(\frac{dv}{dP}) \quad \alpha \quad (D\text{-}4)\log(P) \tag{5}$$

 This shows that the fractal dimension of a porous stone can be obtained from the slope of the plot of pore-size distribution log(dv/dP) vs log(P).
 Figs. 2 and 3 show the pore-size distribution of the limestone beds of the Great Sphinx plotted in the form of log(dv/dP) vs log(P). Fig. 2 shows a typical plot in the range 0.1 - 200 MPa (15 - 30000 psia). There, it can be seen, that the left portion of the curve, 0.1 to 0.7 MPa, does not exhibit a power law relationship and cannot be described by Eq 5. However, for this stratum, beyond 0.7 MPa (100 psia) a power law relationship holds and a fractal dimension can be calculated

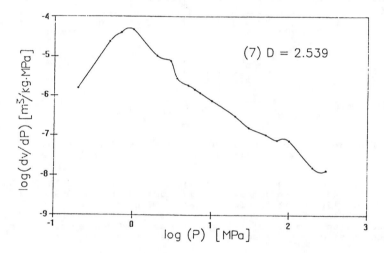

FIG. 2--Pore-size distribution of bed 7 showing the left
portion of the curve that behaves non-fractally
and the fractally behaving right portion.

from Eq 5. This behavior is true of the data for all strata of the Sphinx with the onset of fractal behavior occuring at successively increasing pressure for lower strata as given below.
 Ink-bottle pores are present in the Sphinx limestone. In ink-bottle pores, the large pores are connected to the surface through narrow necks. It has been shown that the mercury porosimetry of rocks with ink-bottle pores does not directly reveal the presence of large pores because, in the intrusion process, these pores are filled at pressures corresponding to the break-through pressure of the connecting necks. Instead, these pores are revealed indirectly through the retention of mercury when the mercury is extruded in the depressurizing mode (Wardlaw 1976, Wardlaw and McKellar 1981). The distribution of these pores, consequently their fractal behavior, is not expressed by the left portion of the curve in Fig. 2. In this figure and in Fig. 3, we show the straight line fitting of the data points for the right hand portion of the curves which clearly obey the power law.

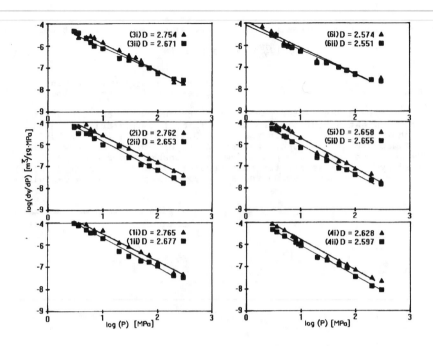

FIG. 3--The plots log(dv/dp) of beds 1-6 of the Sphinx
whence the fractal dimension, D, was calculated.

The D values obtained, following Eq 5, range from 2.54 for bed 7 to 2.77 for bed 1i. The corresponding volume for pores <0.9 μm (0.7 MPa) for bed 7 is 19% and it is 61% for bed 1i with pores <0.18 μm (3.4 MPa). We see that the D values regularly decrease for bed 1 to bed 7. Also, each bed, except bed 7, is distinguished by two D values as the upper and lower bed, with the lower bed characterized by the larger D value.

DURABILITY FACTOR AND FRACTAL DIMENSION

In order to develop a correlation between the durability factors and the fractal dimensions, four points from Table 1 were chosen (beds 7,6ii,2ii, and 2i) so as to cover the full range of D. The expression which fits the data best has a quadratic form and is given as

$$DF = a_0 + a_1 D + a_2 D^2 \qquad (6)$$

where
a_0, a_1, a_2 = constants with values -12100, 9571.41, and -1878.18.

TABLE 1--Durability factors (DF), small pore volume
percentages (V_{sp}), and the calculated fractal
dimensions (D) of the Sphinx limestone.

Bed No.	DF_1^*	DF_2^*	V_{sp}	D
7	100.0	94.1	19	2.54
6ii	98.9	94.3	21	2.55
6i	95.8	93.3	23	2.57
5ii	92.5	74.7	38	2.65
5i	84.9	70.7	37	2.66
4ii	85.8	89.2	24	2.60
4i	75.8	81.6	27	2.63
3ii	76.3	66.3	37	2.67
3i	41.9	17.6	42	2.75
2ii	76.9	74.7	32	2.65
2i	10.8	9.9	58	2.76
1ii	56.4	61.5	38	2.68
1i	11.7	1.7	61	2.77
Indiana Limestone	94.5	84.5	26	2.62

$^*DF_1$: derived from pore-size distribution (Punuru et al. 1990).
DF_2: derived from fractal dimension.

   Fig. 4 shows the plot of the fractal dimension versus durability
factors and volumes of small pores (<0.5 μm). The durability factor

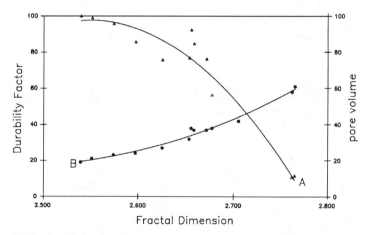

FIG. 4--Plot showing the correlation of the fractal
dimension with: (A) durability factors, and
(B) small pore volumes (<0.5 μm).

values correspond to the data published by Punuru et al. (1990). Curve A
shows the quadratic fit. It can be seen from this figure that though the
regression line has been obtained based on only four points, the rest of
the points show reasonably good agreement. The regression fit of this
curve gives a correlation coefficient value of 0.97. The regression fit
of fractal dimension and small pore volume (Curve B) has a correlation
coefficient value 0.99.

Eq 6 has also been used to determine the durability factor of the Indiana Limestone. The values obtained match the known durability factors of the Sphinx as well as that of the Indiana Limestone (Table 1).

DISCUSSION

We have followed established practice for the selection of pore-size regime in which the Sphinx strata behave fractally (Friesen and Mikula 1987, Avnir et al. 1983, 1984, Kolb 1990, Turcotte 1991). According to this practice, the pore-size distribution in the selected regime should follow a power law, the D values should fall within the range of 2 and 3, and that these values increase with increasing microporosity.

Fig. 2 represents a pattern of pore-size distributions seen in all Sphinx strata for a pressure range of 0.1 - 200 MPa. The left portion of the curve does not obey a power law and thus must be excluded from consideration of fractal analysis. We have explained earlier that this is an artifact of mercury porosimetry that is unable to directly reveal large pore-size distributions in an ink-bottle pore system. The upper limit for fractally behaving pore radii is 0.9 μm in bed 7 and decreases successively to 0.18 μm for the lowermost bed showing that the lower beds have the large pores connected through narrower capillaries than the upper beds where the capillaries are somewhat larger.

Concerning the increase of D-values in increasingly microporous strata, the fractal dimension of volumetric objects is modelled on the basis of roughness of the surface produced due to the presence of pores. Thus, the presence of a large number of small pores will produce many sites of roughness producing a larger fractal dimension for microporous beds.

In conclusion, this study shows that the fractal dimension is a highly sensitive technique to predict rock durability. It may be used to differentiate rock behavior better than can be done by other existing means for allied rocks where other distinctions may not be so obvious.

ACKNOWLEDGEMENT

This research was supported by funds from the Samuel H. Kress Foundation. We are also grateful to our colleagues Profs. John Sinai and Shi Yu Wu who participated in discussions on the subject. Prof. Sinai also helped with the preparation of this manuscript.

REFERENCES

ASTM Standard C-666, 1973, "Soundness of Aggregate by use of Sodium Sulfate or Magnesium Sulfate," Annual Book of ASTM Standards, pt. 10, pp. 49-53.

Avnir, D., Farin, D., and Pfeifer, P., 1984, "Chemistry in Noninteger Dimensions Between Two and Three. I. Fractal Theory of Heterogeneous Surfaces," Journal of Chemical Physics, Vol. 79, pp. 3558-3565.

Avnir, D., Farin, D., and Pfeifer, P., 1984, "Molecular Fractal Surfaces," Nature (London), Vol. 308, pp. 261-263.

Friesen, W.I. and Mikula, R.J., 1987, "Fractal Dimension of Coal Particles," Journal of Colloid and Interfacial Science, Vol. 120, pp. 263-271.

Gauri, K.L., 1984, "Deterioration of the Stone of the Great Sphinx," American Research Center in Egypt, Newsletter, Vol. 114, pp. 19-27.

Gauri, K.L. and Punuru, A.R., 1989, "Characterization and Durability of Limestones Determined through Mercury Intrusion Porosimetry," Proceedings, First International Symposium on The Conservation of

Monuments in the Mediterranean Basin, Bari, Italy, F. Zezza ed.,
    pp. 227-233.
Gauri, K.L. and Yerrapragada, S.S., "Pore-Potential and Durability of
    Limestones," Proceedings, Spring '92 Meeting of Materials Research
    Society, San Fransisco, USA, in press.
Kessler, D.W. and Anderson, R.E., 1951, "Stone Exposure Test Wall,"
    Report 125, p 41, U.S. Dept. Commerce, NBS, Building Materials and
    Structures.
Kolb, M., 1990, "Fractal Characterization of Porous Media: A Model
    Calculation," Journal of Non-Crystalline Solids, Vol. 121,
    pp. 227-233.
Punuru, A.R., Kulshreshta, N.P., Chowdhury, A.N., and Gauri, K.L., 1990,
    "Control of Porosity on Durability of Limestone at the Great
    Sphinx, Egypt," Environmetal Geology and Water Science, Vol. 3,
    pp. 225-232.
Turcotte, D.L., 1991, "Fractals in Geology: What are They and What are
    They Good For?," GSA Today, Vol. 1, pp. 1-4.
Wardlaw, N.C., 1976, "Pore Geometry of Carbonates as Revealed by Pore
    Casts and Capillary Pressure Data," American Association of
    Petroleum Geologists Bulletin, Vol. 60, pp. 245-247.
Wardlaw, N.C. and McKellar, M., 1981, "Mercury Porosimetry and
    Interpretation of Pore Geometry in Sedimentary Rocks and Artificial
    Models," Powder Technology, Vol. 29, ,pp. 127-143.

John-Paul Latham,[1]

A MILL ABRASION TEST FOR WEAR RESISTANCE OF ARMOUR STONE

REFERENCE: Latham, J.-P., **"A Mill Abrasion Test for Wear Resistance of Armour Stone,"** Rock for Erosion Control, ASTM STP 1177, Charles H. McElroy and David A. Lienhart, Eds., American Society for Testing and Materials, Philadelphia, 1993.

ABSTRACT: The motives and historical background for developing a new test for the abrasion resistance of armourstone are described. A brief and critical account of several abrasion tests often used for rock and rock aggregate quality specification is given. The development of a new mill abrasion apparatus and test procedure used at Queen Mary and Westfield College, which overcomes some of the main shortcomings of other tests, is explained.

The abrasion mill test uses increments of milling time to generate a weight loss curve, the slope of which gives the abrasion resistance index. This new index, k_s is compared with wet attrition and aggregate abrasion values for a suite of 11 rocks with wide ranging strengths. Los Angeles Abrasion Test results are briefly mentioned. The weight loss plots and values of k_S from many rock samples are presented showing highly discriminating and reproducible results over the whole range from soft chalk to a tough gneiss. The plots give an immediate physical interpretation of likely comparative degradation rates.

Prediction of degradation rates from mill abrasion test results, is discussed briefly. This prediction uses a previously published degradation model for armourstone that takes account of site conditions.

KEYWORDS: armourstone, abrasion, aggregates tests, Deval test, Los Angeles test, degradation, wear, test correlation, in-service degradation

Wear resistance is an important intrinsic property of rock used in the cover layers of rubble mound erosion control structures. It is a property of the mineral fabric strength of the intact rock. Given exactly the same site conditions of the placed armourstone, it is possible that marginal quality rock types that may have to be considered for use in some parts of the world, could wear away through principally an abrasive action, at rates hundreds or thousands of times faster than more typically excellent rock types. Alternatively, the same rock type could be used in different armour designs and/or exposed to a range in the environmental aggressiveness such that the annual rates of material removal also differ by perhaps hundreds. A test method for wear resistance therefore ideally needs to be sensitive at both weak and

---

[1] Lecturer, Geomaterials Unit, Queen Mary and Westfield College, London University, Mile End Road, London E1 4NS, UK.

strong ends of the scale in order that results can be interpreted in a
wide range of site conditions and that marginal materials can be
evaluated.
   There are a great many test methods that examine hardness and
resistance to abrasion and attrition of rock materials, see for example
the ISRM suggested methods (Brown, 1981). Most tests are intended to
evaluate the performance of aggregates and building materials and some
of these are described later. None have previously been devised for
armourstone, although the ASTM Test Method for Resistance to Abrasion of
Large-size Coarse Aggregate by use of the Los Angeles Machine (C 535) is
frequently proposed for this purpose in America and many other parts of
the world.
   This paper makes reference to an abrasion test called the
Determination of the Abrasion Resistance Index Using the Queen Mary and
Westfield (QMW) Abrasion Mill for which the method was first described
in 1991 (Latham, 1991; CIRIA/CUR, 1991). Interim results from an on-
going programme of research, aimed at further evaluating the potential
of the new test method, will be presented.

BACKGROUND

   The mill abrasion test is of interest as both a performance
simulation and as a more fundamental measure of wear resistance,
although these two objectives are not entirely compatible. The
historical motivation for the test development is described below.
   Interest in this type of mill tumbling test began with
dissatisfaction in the predictive capabilities and test duration of
existing accelerated weathering tests. Another form of simulation test,
one based on the wearing away of particle mass seemed attractive for
durability assessment since rounding and weight loss in this test could
then be compared directly with rounding and perhaps also with weight
loss of armour blocks in service.
   The first test set-up used several 200-300 g blocks of rock which
were abraded against each other by tumbling in a drum rotated about a
horizontal axis which was 10% filled with water. As it became charged
with rock particles, the water required changing at frequent intervals
during the test with some rock types. This introduced an unknown
variable into the test. Nevertheless, comparison of abrasion in this
roller mill and abrasive rounding of the same rock in-situ on a
breakwater proved to be a potentially reliable and sensitive means of
discriminating between rock types with generally similar weathering
resistance characteristics.
   Since that study, improvements in the test apparatus and method,
together with image analysis methods, (Latham and Poole 1987b) have
enabled the relations between weight loss and shape change to be
quantified (Latham and Poole 1988a). These image analysis methods
provided a means of measuring armourstone block shape changes and
rounding (using Fourier shape descriptors) which could then be related
to weight loss predictions. An application of the test to prototype
degradation studies was described from an armourstone revetment in
Scotland (Latham and Poole 1988b).
   A pilot study of the new abrasion mill test was first reported
in 1987, suggesting how an index of abrasion resistance might be
defined. The current test method for the determination of the abrasion
resistance index, using the QMW abrasion mill, has been fully described
elsewhere (Latham, 1991; CIRIA/CUR, 1991). The objective of the test is
to provide good discrimination, over at least two orders of magnitude,
for the material property of the mineral fabric strength that governs

the resistance to mutual attrition of water-saturated coarse aggregates.
The test does not consider the mineral fabric strength to resist
breakage along fresh fractures since this is better achieved by other
intact strength tests such as the fracture toughness test. The wet
conditions of the test will tend to promote any weakening influence of
clay minerals present. The gentle, tumbling-like rolling of shingle on a
beach is a long and progressive process of continually fatiguing the
surface layer of the rock particles.

ABRASION TESTS

(1) The Los Angeles Test (ASTM C 535)

A charge of 12 steel balls, weighing a total of 5 kg, is added to
a rock sample of coarse aggregate also weighing 5 kg which are rotated
in a steel cylinder of 0.7 m diameter for 1000 revolutions at 30 to 33
revolutions per minute. As the rock pieces and balls drop from the
shelf and roll within the drum, whole lump fractures, due to impacts and
crushing, takes place in addition to grinding and abrasive wear. After
sieving out fines (<1.7 mm), the weight loss percentage is the Los
Angeles Value. The test is referred to in ASTM Standard practice for
Evaluation of Rock to be Used for Erosion Control (D 4992) and Lutton
(1988) has estimated an acceptance value of not more than 25% loss in
1000 revolutions as generally appropriate.

(2) Aggregate Abrasion Test (BS 812, 1975)

A grinding lap is applied with a steady load to pads of embedded
aggregate (10-14 mm) for 500 revolutions while sand is continuously fed
in front of the sample. The percentage weight loss is the Aggregate
Abrasion Value. This test method of the British Standard for Testing
Aggregates (BS 812: Part 3, Mechanical properties, 1975) is specified
in the British Standard for Maritime Structures (BS 6349: Part 1:
General Criteria, 1984) where rocks for use as armourstone are required
for acceptance to have a value of not more than 15%.

(3) Wet Attrition (Deval) Test (BS 812, 1951)

The test sometimes known as the Deval test, is performed in a
machine with two cylindrical containers mounted on a shaft at an angle
of 30° with the axis of rotation. The Deval machine is used
internationally and is favoured for testing railway track ballast.
Slight variations in test method for the French, (AFNOR norme Essai
Deval humide NF P 18-577), British (BS 812) and American (ASTM D2-33)
standards exist. The French have probably the most experience with this
test and employ it as a current standard for armourstone. The UK and
USA standards have been withdrawn. However, this type of test, using
the Deval machine, is currently the most favoured of existing standard
tests now under consideration for the Eurostandard for stone in
hydraulic structures for assessing resistance to wear. For the wet
attrition test, the aggregate pieces are typically 37 to 50 mm sieved
lumps and the test sample of 5 kg is placed in the cylinder to which 5
litres of clean water is added. The test is run for 10,000 revolutions
(5 hours) and the weight retained on the 2.36 mm sieve is used to
calculate the percentage weight loss. Requirements for the French test
(which differs slightly in detail from the BS wet attrition test), for
use in armourstone specifications are set at values equating roughly

with <8%, <10% and <13% for percentage wear for the three different
quality classes (A, B and C) of armourstone (LCPC 1989).

(4) Micro-Deval Test (AFNOR norme Granulats - Essai d'usure micro-Deval NF P 18-572)

This test may be run dry or with 2.5 litres of water added to a
sample of 10-14 mm aggregates weighing 500 g and a charge of 5 kg of 10
mm steel balls. These are rotated in 200 mm diameter drums at 100
revolutions per minute about a horizontal axis for 12,000 revolutions (2
hours) and the percentage weight loss is determined on a 1.6 mm sieve.
A modification to the test, currently under consideration for the
Eurostandard for stone in hydraulic structures, may permit a test sample
of 25 to 50 mm to be used in the micro-Deval test.

(5) Sandblast Test

The sandblast test proposed by Verhoef (1987) propels quartz sand
under air pressure at rock specimens. The sandblast index expresses the
ratio of the volume loss of the specimen with that of soda-lime glass
(as a reference material). The test's mechanism of wear, which is
measured by repeated impacting causing tensile failure spalling on the
grain scale, suggests a strong correlation with tensile strength would
be expected. This is what is found. The sandblast index is extremely
nonlinearly related to tensile strength giving very good discrimination
for weaker rocks.

DISCUSSION OF ABRASION TESTS

Some disadvantages with these tests are now discussed. The tests
on aggregate samples that involve considerable impacting (tests 1 and 4)
are entirely suitable for rock to be used as aggregate of the same size
as the test aggregate. But where large rock is concerned, the need for
preparation of test aggregate from representative pieces of armourstone
can introduce variations due to preparation methods that may be
significant. Test sample material crushed through quarry processing can
also vary in quality as a function of that processing. Coarse grained
igneous rock often performs very poorly on any tests involving impacts
where the aggregate size is relatively small. The mineral fabric
strength to resist impact breakage can be assessed using an intact
strength test such as fracture toughness (ISRM 1988) or more simply by
the point load strength index (ISRM 1985). Tests which combine
impacting with surface grinding do not allow an evaluation of wear
resistance that is assessed independently of toughness. This means that
the rock properties can not be as ideally matched with the intended site
application.
The abrasive simulation in test 2, designed for highway
applications, is the least representative for armourstone and as with
test 5, it does not detect any water weakening effects important for
hydraulic applications.
The weaker rocks in tests 1, 3 and 4 are cushioned (sometimes in a
viscous slurry for test 3) or actively abraded by the finer material
during the test (which in practice would be washed out). The abrasive
forces are not constant during this test. At an arbitrary test
completion time, this gives little more than a ranking of wear
resistance and usually a poor discrimination between weak rock types.

MILL ABRASION TEST

Method summary

    A narrow size range of aggregate, sieved to between 26.5 and 31.5 mm is prepared. Two mills (see Fig. 1) are half filled with equal weight test samples (usually about 4.5 kg depending on density and packing) by dividing a filled cylinder. No abrasive charge is added and axially mounted ribs prevent internal sliding and promote tumbling. Incremental milling at 33 revs. per minute, with a continuous water feed passing through a 14 mm diameter axis hole, is performed. One of three incremental programmes is selected for the test depending on whether the rock is determined to be weak, intermediate or strong as defined by the weight loss measured after 1000 revs. Ultra-violet tracer lumps are added to make up weight loss after each increment and only the saturated surface dry original lumps are used to monitor the weight change. The test is stopped when the fraction of the original weight remaining, W/W_0 has fallen below 0.7 or after 200,000 revs. The graphs are plotted and the abrasion resistance index, k_S is calculated as shown in Fig.2. This approach to evaluating the significance of the weight reduction plot was based on theoretical considerations (Latham and Poole 1987a, 1988a).

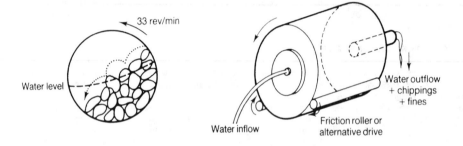

33 rev/min

Water level

Water inflow

Water outflow
+ chippings
+ fines

Friction roller or
alternative drive

Mill specifications: Inside diameter 195 mm, inside length 230 mm
Outflow diameter 14 mm
Three plastic lifting ribs of 20 mm diameter, semicircular
section, equally spaced and axially attached to inside

FIG. 1 The QMW abrasion mill (schematic).

Method Advantages

    (i) The abrasive environment is gentle enough not to exploit cracks artificially introduced to differing degrees using different test aggregate preparation methods. This is certainly not the case for other aggregate tests with steel balls in the mill which tend to measure toughness more than abrasion resistance (e.g. micro-Deval and Los Angeles).
    (ii) The abrasive environment remains approximately constant during milling so that all nonlinear features of the fractional weight loss versus time plot can be attributed to material behaviour rather than artifacts of the test process. Abrasion resistance index results can therefore discriminate just as well at the high and low quality

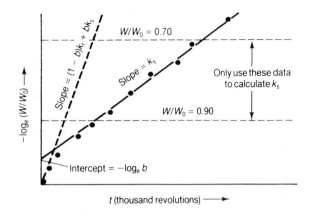

FIG. 2 Determination of the abrasion resistance index, k_S.

ends of the performance range doing more than just ranking rock types. This is achieved by two procedural features of the test.

(a) Chippings and most fines are removed continually as a suspension in water so that a viscous slurry or abrasive charge of fragments cannot build up. This build-up often happens to varying unknown degrees in the Deval test which is run for one set time increment only.
(b) Fresh material (identified by ultra-violet tracer) is added to make up the weight lost after each increment of milling so that with an initially half-filled cylindrical mill of similar sized pieces, the number of particle diameters rolled across the diametral plane per mill revolution is reasonably constant throughout the test.

(iii) The abrasion resistance index is calculated from the slope of the progressive weight loss curve between 0.9 and 0.7 of its original weight. This helps to remove any differences in the initial weight loss regime that are mainly attributable to initial shape, (see Fig. 3). In any case, a limit to the amount of flaky pieces is imposed.
(iv) One of the most important features of the test is that the results yield a fractional weight loss versus time plot resulting in an easily visualized model for interpretation of in-service durability of different rock types.
(v) The procedure is aimed at giving high accuracy over a wide range of rock properties. This is in part achieved by the narrow size range of the sample. Repeatability and reproducibility results are not available. However, the six plots shown in Fig. 3 give some indications of what could be expected from a relatively homogeneous rock source such as Carboniferous limestone. The values calculated for k_S are given in Table 1. They indicate that the slightly faster rotation rates and the increased mill diameter of the new mill versus the old mill contributed to a relative factor of 1.50. This in turn relates old mill to new mill k_S test results. A 50% proportion of flaky shaped material resulted in only a 5% reduction of k_S. This is because any

influence of initial shape is associated more with the slope of the
early part of the weight loss curve, as indicated by k_f, than with k_s
(see Fig. 2).

TABLE 1   Calibration data to assess flakiness and new/old test
procedure.

Test Condition	Abrasion Resistance Index, ks
New A (0% flaky)	0.002549
New B (0% flaky)	0.002573
Old A (0% flaky)	0.001657
Old B (0% flaky)	0.001762
New A (25% flaky)	0.002513
New B (50% flaky)	0.002377

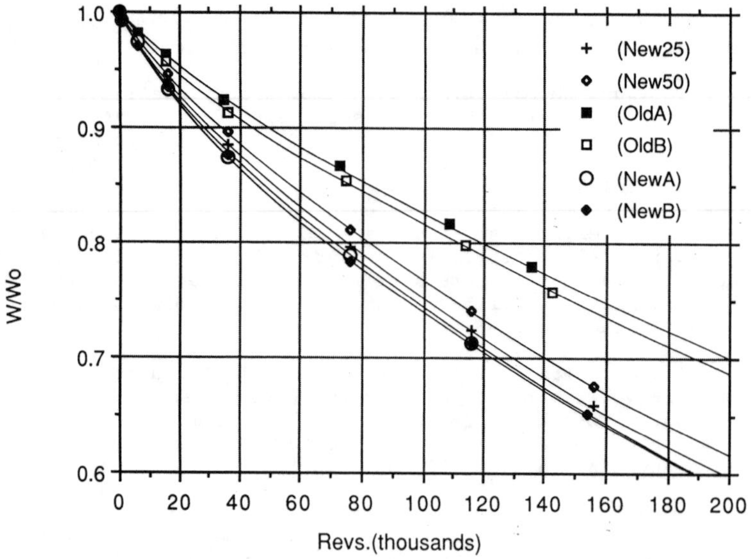

FIG. 3   Mill abrasion test plots of six tests on the same rock (see
text).

Method disadvantages

(i) Limited experience for the test (and apparatus not widely
available, although commercial production of a suitable apparatus is
under negotiation).

(ii) Possible lengthy sample preparation because of narrow size
range.

(iii) Test duration takes up to 2 weeks for reporting on excellent
rocks and requires more labour time (about 5 hours) than simple one
increment tests.

(iv) Careful work practices are required to ensure test material is correctly accounted for at each increment.

(v) It could be argued that the test does not simulate real wear processes as well as other tests which involve impacting to a greater degree. The wear simulation, which is of mutual grinding and attrition, is like that of rocking abrasion on site but at much lower stresses. It is rather unlike the pounding by shingle or blasting by sand with a uniform hardness typical of quartz particles.

RESULTS

Data for a comparison between 11 rock samples and three tests is presented in Table 2 and plotted in histogram form in Fig. 4.

TABLE 2   <u>Abrasion resistance index, k_S, compared with other abrasion test results</u>

Rock type and sample code		Mill Abrasion Test k_S	Wet Attrition Test WAV	Aggregate Abrasion Test AAV
sandstone	fs	0.01470	24.2	15.71
sandstone	bs	0.02450	27.4	42.91
hard chalk	mc	0.01790	29.1	29.64
calcareous sst	ob	0.01100	24.1	15.14
sandstone	ts	0.00083	4.3	2.70
granite	hg	0.00069	3.5	2.69
limestone	or	0.00405	16.6	8.94
oolitic lmst	no	0.00703	15.9	14.95
dolerite wthd	tdw	0.00778	22.0	8.46
granite s.wth	hgw	0.00135	7.0	4.58
dolerite	hd	0.00286	8.0	5.91

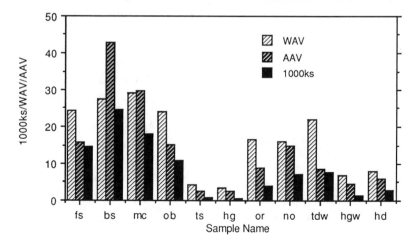

Fig. 4   Bar chart comparison of test results

The curves plotted in Fig. 5 and Fig. 6 are derived from the following theoretical equation that describes the weight loss in terms of a smoothing resistance index k_f, the abrasion resistance k_s, and a mixing term b, (Latham and Poole 1988):

$$W/W_0 = (1-b) \exp(-k_f t) + b \exp(-k_s t)$$

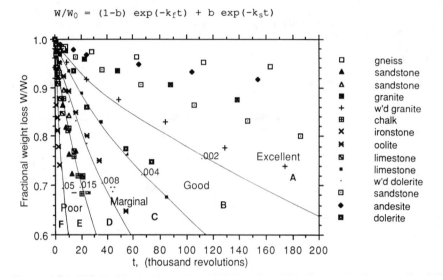

FIG. 5   A selection of plots from the mill abrasion test.

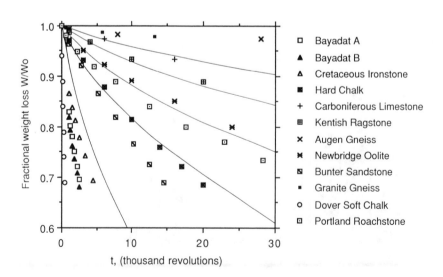

Fig. 6   Mill abrasion test plots emphasising poor and marginal rocks.

Values of k_f/k_s = 30 and b = 0.95, have been substituted together with k_s values of 0.002, 0.004, 0.008, 0.015 and 0.05 to produce the curves which roughly classify the plot field into zones A to E. For more general rock quality classification purposes, a four-fold division into excellent, good, marginal and poor has been proposed.

It is interesting to note that tests on tough rocks can reach the cut-off duration of 200,000 revolutions while weights have not even dropped by 10%, giving k_s values as low as 0.0002. The same test, on a range of weak chalks, could be performed on the weakest material. Here, milling increments of 100 revolutions leading to test completion of over 30% weight loss can occur after less than 1000 revolutions yielding a k_s value of about 0.7. Consideration and testing of such weak rocks may seem irrelevant if poor rocks need never be used. Fig. 6 includes some plots from very poor materials. The Bayadat limestone is a duricrust limestone used for many breakwaters in the Persian Gulf where the wave conditions and abrasive environment are exceptionally mild.

The method for calculating k_s by a least squares regression is shown in Fig. 7 for one sample of Bayadat limestone which was divided and tested in the two mill cylinders A and B. This gives an indication of good repeatability even for inhomogeneous weak materials. The coefficient of variation, expressed as the standard deviation divided by the mean of k_s is probably within 5% for a range of k_s from 0.1 to 0.0001. The mill abrasion test would therefore appear to be very useful for discriminating the best sources of locally available duricrust limestones. Other methods for selecting from a choice of poor and marginal rock types were described by Fookes and Thomas (1986).

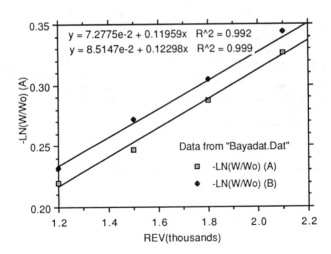

Fig.7 Least squares regression to determine the abrasion resistance index, k_s. For cylinder A: k_s = 0.123, cylinder B: k_s = 0.120.

CORRELATIONS

$\underline{k_s \text{ versus WAV}}$

The data from Table 2 has been plotted in Fig. 8 and Fig. 9. The correlation coefficient is good, which is to be expected since both tests are of mutual attrition by similar mechanisms. It is relatively simple to predict W/W_0 at 10,000 revs. in the abrasion mill for a given k_s value using the above equation (with $k_f/k_s = 30$ and $b = 0.95$) or from Fig. 6. The equivalent % loss in weight

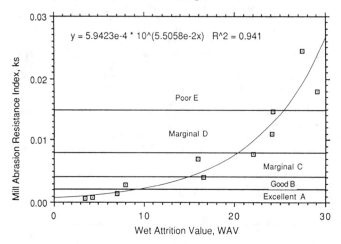

Fig.8   Mill abrasion resistance index versus wet attrition value.

given by the WAV for $k_s = 0.001, 0.005, 0.01$ and 0.03 is respectively approximately a factor of 3, 1.8, 1.5 and 1.0 times the value predicted for inside the abrasion mill. The nonlinear relationship can be easily explained. For excellent quality rocks, 10,000 revs. will have only rounded the sharp corners and the total energy imparted to each piece per revolution in the Deval machine has remained roughly constant because there has been minimal loss of weight per lump. For a rock type with $k_s = 0.03$, the total energy imparted has dropped considerably after

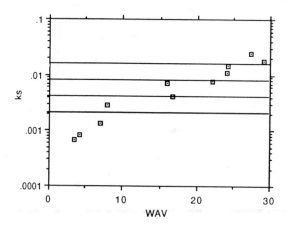

Fig. 9   Data as for Fig. 8 plotted with log scale.

10,000 revs. as the pieces have reduced in weight by about 30%. This may also be attributed in part to the formation of a viscous slurry and other nonlinear factors mostly removed from the mill abrasion testing procedure.

Fig. 9 shows the same results as Fig. 8 on a log scale. This limited set of results suggests that both k_s and WAV can distinguish excellent rocks perhaps equally well but this may not be so for good and marginal rocks which show quite significant differences if the best-fit equation is used to predict k_s, (see for example rock types hd, no, or and tdw). Examination of the log plot shows that there is perhaps a case to be made for dividing the excellent class into two zones at a value of k_s = 0.001, which would remove the slightly weathered granite and Carboniferous limestone from the very best category. This would give three top quality bands; excellent, extremely good, and good for k_s boundaries at 0.001, 0.002 and 0.004 conforming closely with the three class boundaries given for the French specifications A, B and C in their wet Deval test.

k_s versus AAV

Data from Table 2, plotted in Fig. 10 shows a linear relation overall but this could not be described as a good correlation. The best-fit equation would give misleading predictions of k_s (compare no, ob and fs, each having an AAV of about 15 but a widely differing k_s). The BS acceptance value of 15% for AAV is near the middle of the marginal quality k_s values, a value of 10% seeming more appropriate to exclude marginal abrasion resistance. Note also the anomalously good performance of the weathered dolerite in the AAV test.

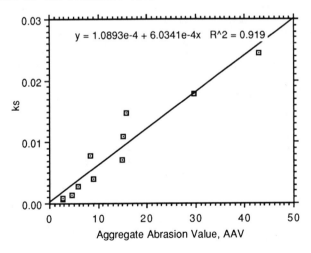

Fig. 10   Mill Abrasion Resistance Index versus Aggregate Abrasion Value.

k_s versus LAV

A small data set comparing Los Angeles test values for 1000 revs. with k_s results is shown in Fig. 11. Again the relationship is nonlinear for the same reasons as described above for the wet attrition test. Unfortunately these data are from a different earlier sample suite. Although they cannot be taken as representative, they do indicate that remarkably poor correlations can be expected. Some comments about specific samples tested are as follows.

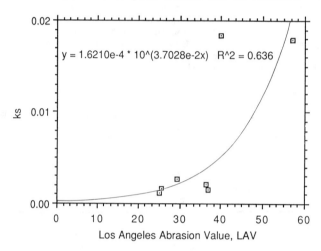

Fig. 11 Poor correlation between k_s and LAV (1000 revs.) from few data.

Two tough coarse grained granites with LAV of 25% have apparently suffered some internal cracking during sample preparation with a jaw crusher and are giving higher LAV results than expected. A brittle pure micritic limestone gave a LAV of 37% while performing well in the mill test giving a k_s of 0.0015. A silty limestone giving a 40% LAV was extremely susceptible to water weakening during the mill test giving a k_s value of only 0.018. The LAV test may not indicate water weakening and is obviously more a test of impact strength than of surface wear.

DEGRADATION PREDICTION

It should be recalled that one motivation for the abrasion mill test development was to use results to give quantitative predictions of in-service degradation. It is not the intention to fully describe the degradation model here as it has been described with illustrative examples elsewhere (Latham, 1991; CIRIA/CUR, 1991) but the principles are briefly discussed.

The aggressiveness of the site and armourstone application is assessed using a system of ratings. Each of nine relevant parameters and numbered X_1 to X_9 and are ascribed a rating value using a crudely calibrated guide table of ratings (Latham, 1991). (These parameters can be grouped into production influenced geometric factors, environmental boundary conditions, and design factors). The ratings can then be multiplied together to give a single factor X, which equals *the number of years in service divided by the equivalent number of thousands of revolutions in the mill abrasion test,* called the *equivalent wear time factor.*

The mill abrasion test plots such as those in Fig. 5 and Fig. 6 can then be mapped into fractional weight reduction of armourstone

against years on a structure by setting different X (i.e. site factor)
values for converting revolutions (in thousands) to years. It may be
noted that although it would seem that of the intrinsic rock properties,
only the abrasion resistance has been included in the degradation model,
this is in fact not so as several site parameters given in the ratings
guide table (wave energy, X_4 and meteorological climate, X_6) link the
intrinsic properties of integrity of blocks and weathering resistance to
the severity of site conditions expected. However, this comparison of
rates of percentage weight loss in years for different rock types and
site conditions gives a prediction of in-service weight losses and
rounding which is perhaps biased more towards abrasive wear than
cracking deterioration.

From the points on the curves in Fig. 5 corresponding to a value
of $W/W_o = 0.9$ - (a very serious degree of prototype degradation probably
equivalent to failure) - it is possible to deduce an approximate
physical interpretation for different ranges of X values if the
degradation model is invoked. For example, for $k_s = 0.002$ and X = 2 to
10, a 10% weight loss occurs at the very soonest, between 60 and 300
years but could take much longer for the large number of fresh rocks
with k_s less than 0.001. Other figures for different combinations of k_s
and X lead to different in-service durability. Armourstones with
properties as poor as Bayadat limestone would need to be applied to
conditions in which X was at least 25 in order to survive less than 10%
weight loss in 25 years service according to this degradation model.

It must be pointed out that the model should not be considered as
fully operational, but rather as an additional design tool.

INFLUENCE OF DURABILITY ON DESIGN OPTIONS

Sometimes new quarries have to be opened with the best rock that
is locally available or a selection of less than ideal sources must be
considered. If there is a convenient local source but of only marginal
quality material, a number of options will need to be considered at the
design stage. These might include:
(i) the high frequency of maintenance and repairs;
(ii) over-dimensioning of armourstone;
(iii) gentler seaward slopes and greater material volumes:
(iv) the use of higher quality rock from a more remote source.
The in-service durability assessment adopting the degradation model
concept may provide a basis to address these whole-life cost
optimization decisions. Sometimes a change in design is called for when
different density material is proposed from that of the design
calculations. Similarly, available rock quality may call for different
designs to be considered.

CONCLUSIONS

The new mill abrasion test method is concerned with the property
of resistance to surface grinding wear in wet conditions and not impact
strength that can be specifically measured with intact rock tests.
Other tests, with the exception of the Deval type or wet attrition test,
tend to measure a hybrid of resistances to abrasion generally dominated
by a different impact-dominated wear mechanism.

The mill abrasion test procedure has the advantage of being
linearized in such a way that abrasion resistance index results will do
more than rank rock types in terms of performance. Rock A with an index
differing by a factor of 100 from rock B may be expected to suffer the
same initial weight losses by grinding but at about 100 times the rate.

It can be used identically on soft chalks and tough gneiss. This is
likely to be important for marginal materials and evaluation of in-
service degradation since the site environment aggressiveness can vary
considerably.
    The test data so far indicates good repeatability although further
test evaluation is considered necessary.

REFERENCES

Brown, E. T., 1981, Rock Characterisation Testing and Monitoring -
    International Society of Rock Mechanics Suggested Methods,
    Pergammon Press, Oxford.

CIRIA/CUR, 1991, Manual for the Use of Rock in Coastal and Shoreline
    Engineering. Published jointly by the Construction Industry
    Research and Information Association in the UK, (Special
    Publication 83) and the Centre for Civil Engineering Research,
    Codes and Specifications (Report 154) in the Netherlands.

Fookes, P. G. and Thomas, R. S., 1986, "Rapid site appraisal of
    potential breakwater rock at Qeshm, Iran," Proceedings of the
    Institution of Civil Engineers, Part I, Vol. 80, pp 1297-1325.

ISRM, 1988, International Society of Rock Mechanics. "Commission on
    Testing Methods. Suggested method for determining the fracture
    toughness of rock," International Journal of Rock Mechanics
    Mining Science and Geomechanics Abstracts Vol. 25, pp 71-96.

ISRM, 1985, International Society of Rock Mechanics. "Commission on
    Testing Methods. Suggested method for determining Point Load
    Strength (revised version)," International Journal of Rock
    Mechanics Mining Science and Geomechanics Abstracts Vol. 22, pp
    51-60.

Latham, J-P., 1991, "Rock degradation model for armourstone in coastal
    engineering". Quarterly Journal of Engineering Geology, London
    Vol 24, pp 101-118.

Latham, J-P. & Poole, A. B. 1987a, "Pilot study of an aggregate abrasion
    test for breakwater armourstone," Quarterly Journal of Engineering
    Geology, London Vol. 20, pp 297-310.

Latham, J-P. and Poole, A. B. 1987b, "The application of shape
    descriptor analysis to the study of aggregate wear," Quarterly
    Journal of Engineering Geology, London Vol 20, pp 311-316.

Latham, J-P. and Poole, A. B. 1988a, "Abrasion testing and armourstone
    degradation," Coastal Engineering, Vol. 12, pp 233-255.

Latham, J-P. and Poole, A. B. 1988b, "Assessing the effect of
    armourstone shape and wear," Proceedings 21st International
    Conference on Coastal Engineering, Malaga, Spain, pp 2299-2312.

LCPC, 1989, Les Enrochments, Ministere de l'equipment, du logement, et
    de la mere. Laboratoire Central des Ponts et Chaussees.

Lutton, R. J. 1988, "Material characteristics of large stone in American construction practice," <u>Transactions of the Society for Mining, Metallurgy and Exploration</u> Vol. 286.

Verhoef, P. N. W., 1987, "Sandblast testing of rock," <u>International Journal of Rock Mechanics Mining Science and Geomechanics Abstracts</u> Vol. 24, pp 185-192.

Henry H. Fisher[1]

INSOLUBLE RESIDUE OF CARBONATE ROCK AND ITS APPLICATION TO THE
DURABILITY ASSESSMENT OF ROCK RIPRAP

REFERENCE: Fisher, H. H., "Insoluble Residue of Carbonate Rock and its
Application to the Durability Assessment of Rock Riprap," Rock for
Erosion Control, ASTM STP 1177, Charles H. McElroy and David A. Lienhart,
Eds., American Society for Testing and Materials, Philadelphia, 1993.

ABSTRACT: Measurement and identification of insoluble residues is a
technique that can be used to estimate the resistance of carbonate rocks
to weathering. Insoluble residue test results are presented for
specimens of quarried rock from several locations in the midwest and are
compared to results from sodium sulfate soundness tests. Rocks
containing insoluble residues greater than 20% clay also have high
sodium sulfate soundness loss and normally poor durability under surface
weathering conditions. This is particularly true for eastern Ohio
limestones formed in a freshwater environment.

KEYWORDS: insoluble residue, rock durability, riprap, soundness, rock
weathering

INTRODUCTION

Insoluble residues of limestones and dolomites are the
noncarbonate materials that remain behind after dissolving the rock in
acid. The most common residues are quartz, chert, clay, or pyrite.

Studies of insoluble residues were made as long ago as 1888
(Krumbein and Pettijohn). They have since been used in stratigraphic
correlation of well cuttings, estimating the durability of carbonate
aggregate for concrete, as well as evaluating the frictional properties
of concrete aggregate used in road pavements (Foster 1991).

---

[1]Geologist, Soil Conservation Service, USDA, 200 North High Street, Room
522, Columbus, Ohio 43215

The purpose of these studies was to investigate the use of insoluble residue testing as an aid in estimating durability of carbonate riprap, particularly on carbonate rocks containing clays. The hypothesis of using insoluble residues as an aid in assessing limestone durability occurred after field examination and laboratory tests made on limestone.

The studies conducted by the author have been primarily confined to Pennsylvanian and Permian freshwater limestones of southeastern Ohio, where these rocks are quarried mainly for roadfill. The limestones crack and break within several years in roadcut exposures. The limestones in question are not used as riprap on Soil Conservation Service (SCS) projects, but are occasionally used for this purpose by municipalities. Studies show these freshwater limestones have a high percentage of clay. The insoluble residue test is not helpful in evaluating Ohio's marine limestones which normally have a clay content of less than 5%. Ohio's marine limestones, which have shale partings that cause deterioration of rock, normally do not have clay contents greater than 5%.

TESTING PROCEDURE

A minimum of five representative samples were taken from each source for insoluble residue testing. Specimens from visually nonhomogeneous rocks (banded, brecciated, or with color changes) were cut perpendicular to the bedding and into thin slabs 2 to 4 mm thick. Random portions of the slabs were used as test specimens. Specimens of visually homogeneous rock were obtained by breaking a mass from the sample. Further crushing of specimens to a size smaller than 4.76 mm (No. 4 sieve) was helpful in promoting solubility in acid.

The mass of each specimen tested was 40 to 50 g. Specimens were oven dried, their masses were recorded, then placed in a large container and dissolved in dilute hydrochloric (muriatic) acid, approximately 1 part acid to 3 parts water. Krumbein and Pettyjohn state that "Commercial HCl (muriatic acid) may be used, for insoluble residue testing, but the formation of gypsum (due to the presence of sulphate ions in the acid) should be guarded against" (Krumbein and Pettyjohn 1938).

Care was taken to prevent the solution from foaming over the container walls, a common problem when a large amount of clay was liberated. In addition to foaming, the acid solution usually darkened as the clay liberated. After foaming stopped, the mixture was transferred to a smaller container. Heating was needed to dissolve some rocks. A rubber-tipped pestle was used to crush any agglomeration that prevented rock from dissolving. After the rock was completely dissolved and the residue settled, the acid was decanted. The remaining solution with residue was filtered through a previously weighed medium-fast filter paper. The filter paper, with residue, was washed with water to remove any remaining acid, oven dried, and weighed.

The insoluble residue is expressed as the mass of residue divided by the mass of original rock, multiplied by 100, and reported to one decimal place.

$$\% \text{ Insoluble Residue} = \frac{\text{Oven Dry Mass of Residue}}{\text{Oven Dry Mass of Specimen}} \times 100$$

Additional tests such as petrographic examination or X-ray can be made on the residue.

RESULTS OF STUDIES

For several years, rock from one quarry in west-central Noble County had been suggested as a source for riprap. It is a gray to dark gray freshwater limestone, homogeneous, dense, showing conchoidal fracture, from the upper part of the Monongahela Group. In 1986, bulk specific gravity and absorption tests were made on six samples. During the soaking in water, one sample split in half and another sample had edges that flaked. One sample had an insoluble residue of 13.9% and five others had residues that varied from 24.6% to 45.3%. In 1987 sodium sulfate soundness and insoluble residue tests were made on samples from this source. The results are presented below.

TABLE 1--Results of tests.

% Soundness Loss	% Insoluble Residue
93.7	45.3
0.8	30.1
70.7	24.6
44.5	not tested
18.3	25.6
4.5	13.9

The average insoluble residue of all specimens tested was 27.9%; the average soundness loss was 38.8%. Only two specimens had soundness losses less than 10%; and the average insoluble residue of just those two specimens was more than 20%. All except these two examples would fail a soundness test that permits a maximum loss of 10%. Petrographic examination indicated that the insoluble residue was clay.

Rock used as riprap within a rock chute (on a non-SCS project) was visually examined and estimated to have 10% of the rock cracked within two years of project completion.

Insoluble residue tests were made in 1991 on similar rock from another nearby quarry in western Monroe County. The results compared to soundness loss from slabs are list in Table 2.

TABLE 2--Results of tests.

% Soundness Loss	% Insoluble Residue
99.5	29.3
28.0	22.5
2.4	21.4
9.5	30.7
89.1	20.1

The average of the five insoluble residue tests was 24.8%; the average soundness loss was 45.7%. All specimens showed an insoluble residue greater than 20%, and all except two of these examples would fail a soundness test that permits a maximum loss of 10%. Petrographic examination determined the insoluble residue in this rock to be clay. Sodium sulfate soundness tests, previously made in 1989, on rock from this source yielded losses of 7.5%, 20.2%, 48.2%, and 36.9%.

Rock from the Monroe County quarry has been stockpiled near the plant for "several years," although the exact length of time could not be obtained. The rock is deteriorating and cracking. The insoluble residue of clay varies from 15.3% to 19.0%.

Other examples of test results for Ohio carbonate rocks are listed below. The first source is freshwater, all others are marine in origin. Five or six slabs per source were tested.

TABLE 3--Results of tests.

County	% Soundness Loss	% Insoluble Residue
Muskingum	13.2	78.3
Darke	4.6	8.5
Greene	2.5	18.4
Miami	1.4	1.8
Miami	13.5	22.2

Again, those examples that fail a soundness test which permits a maximum loss of 10% have insoluble residue values greater than 20%.

Another example of nondurable rock containing a high insoluble residue is the freshwater limestone of the Claron Formation in Bryce Canyon National Park, Utah. A combined sample of pale orange and pale pink rock yielded a clay insoluble residue of 27.8%. This limestone erodes readily to produce the picturesque shapes of the park.

An insoluble residue of quartz may not have any effect on rock durability as shown by a specimen from Richland County, Wisconsin. This specimen has a 38.6% insoluble residue of fine quartz, but a soundness loss of only 8.5%.

INTERPRETATION

Dunn and Hudec studied argillaceous limestones from the standpoint of soundness of concrete aggregate (Dunn and Hudec 1966). They found that some rocks disintegrated from simple wetting and drying. Their conclusions were "clays are rejected by dolomite when it replaces calcite in argillaceous limestones. The rejected clay tends to form a continuous network, and thereby becomes wettable." Their studies further indicated that water on clay surfaces is capable of exerting great force on the pores of the rock causing it to disintegrate.

Shakoor, West, and Scholer state that the worst limestone for aggregates, from a durability standpoint, contains more than 20% clay evenly distributed throughout the rock (Shakoor et. al. 1982). Using the 20% figure, nine of the ten samples tested in Tables 1 and 2 would be rejected by the insoluble residue test. Although a direct correlation between insoluble residue and soundness does not exist, both tests would indicate that use of material from both quarries would be highly questionable.

The maximum sodium sulfate soundness loss for riprap permitted under SCS construction specifications is 10% (USDA). Using the soundness test alone, seven of the eleven samples noted in Tables 1 and 2 would be rejected, with one borderline sample; and rejection of two samples in Table 3. All of the samples that would be rejected by insoluble residue would be rejected soundness losses. The insoluble residue test would also reject the borderline sample and the two additional samples. The fact that rock contains a moderately high percentage of insoluble residue of clay is believed by this author to have a bearing on the ability of the rock to withstand destruction in sodium sulfate testing, and to also resist weathering.

CONCLUSION

Based on personal field observations, the author believes that carbonate rock durability containing 20% clay as an insoluble residue must be viewed as highly questionable for use as riprap, and those containing 15 to 20% clay may also be suspect.

The insoluble residue test can be helpful in evaluating carbonates that contain clay. The advantage of this particular test is that it can be run in one to two days, a much shorter time period than needed for some of the other predictive tests such as freeze-thaw or sulfate soundness. The insoluble residue test can be performed in commercial laboratories not equipped for making detailed geological tests, such as X-ray or DTA on clay minerals. In addition, this test can also give a percent composition of insoluble materials within the total mass of rock.

It is the author's opinion that no single test presently exists that will provide a definitive prediction of weathered riprap durability for all types of rocks and no single sample from a source of rock is an adequate quantity on which to base the prediction. Actual long-term weathering on projects is the ultimate test of rock performance. Comparisons of all rock properties and tests including insoluble residue with field observations of weathered riprap would be desirable, and make a useful study.

APPENDIX

In all of the above tests, except for Bryce Canyon and Wisconsin, the sulfate soundness tests were run on 2-1/2 inch (6.4 cm) thick slabs cut perpendicular to the rock bedding, according to ASTM test method D5240 developed within ASTM Subcommittee D18.17 (ASTM 1993). The specimen is soaked and dried in a sulfate solution according to the procedure stated in Section 8 of ASTM test method C88 (ASTM 1992). The percent loss in method D5240 is based on the mass of the largest remaining fragment after testing, then divided by the initial mass of the slab before testing, and multiplied by 100.

$$\% \text{ Soundness loss} = \frac{A-B}{A} \times 100$$

Where A = Oven Dry Mass of Slab Before Testing
      B = Oven Dry Mass of Largest Remaining Piece of Slab
          After Testing

REFERENCES

ASTM, Standard test method for Soundness of Aggregates by Use of Sodium Sulfate or Magnesium Sulfate, C88, *Annual Book of ASTM Standards,* Vol. 04.02, 1992.

ASTM, Standard test method for Testing Rock Slabs to Evaluate Soundness of Riprap by Use of Sodium Sulfate or Magnesium Sulfate, D5240, *Annual Book of ASTM Standards,* Vol. 04.08, 1993.

Dunn, J. R. and Hudec, P. P. Water, Clay and Rock Soundness, *Ohio Jour. of Science,* Vol. 66, March, 1966, pp. 153-168.

Foster, S., Chairman ASTM Subcommittee D04.51, Personal Communication, 1991.

Krumbein, W. C. and Pettijohn F. J., *Manual of Sedimentary Petrography,* D. Appleton-Century Co., 1938.

Shakoor, A., West, T. R., and Scholer, C. F., Physical Characteristics of Some Indiana Argillaceous Carbonates Regarding Their Freeze-Thaw Resistance in Concrete, *Bull. of Assn. Eng. Geologists,* Vol. XIX, No. 4, 1982, pp. 371-384.

USDA Soil Conservation Service, Rock for riprap, Material Specification No. 523, <u>National Engineering Handbook, Section 20.</u>

Dennis M. Duffy[1] and Hilda H. Hatzell[2]

ENVIRONMENTAL TESTING OF ROCK USED AS EROSION PROTECTION IN ARID
ENVIRONMENTS

REFERENCE: Duffy, D. M. and Hatzell, H. H., **"Environmental Testing
of Rock Used as Erosion Protection in Arid Environments,"** Rock for
Erosion Control, ASTM STP 1177, Charles H. McElroy and David A.
Lienhart, Eds., American Society for Testing and Materials,
Philadelphia, 1993.

ABSTRACT: Crushed rock fragments are used in arid environments to
control slope erosion. A test technique was developed to asses the
durability of these fragments when solar heated to summer temperatures
and then "quenched" by thunderstorms containing hail. Aggregates were
heated to 66^0 C and maintained at that temperature for 23 hours. At the
completion of the heating cycle the rock fragments were cooled to room
temperature, over a one hour period, then soaked for 24 hours in water.
The water was decanted and the particles placed back in the oven for an
another 24 hours of heating. This 48 hour cycle was repeated for a
minimum of 50 times. At the completion of the 50 cycle test period the
grain size distributions of the materials were compared to the original
size distributions. Each of the rock types experienced some distress
although most were judged suitable for use as slope protection.
Observation of actual slope protection endurance is continuing for
several of the rock types placed on freeway slopes.

KEYWORDS: Rock fragments, slope erosion, slope protection, arid slope
protection

In arid environments the control of erosion on soil slopes

---

[1] Associate Professor, Department of Civil Engineering, Arizona State
University, Tempe, AZ 85287.

[2] Formerly Assistant Professor, College of Engineering, Arizona
State University, currently with the US Geological Survey in
Tallahassee, FL 32301.

is hampered by the availability and cost of water. The development of eroded surfaces produces both esthetic and operational difficulties. Sediment eroded from the slopes of the freeway, AZ 360 was destroying pumps used to prevent freeway flooding, Figure 1.

figure 1, Erosion of two to one slope on AZ 360 following summer Thunder storm

Vegetation such as grasses, is not a viable erosion mitigation technique without an economical source of water. Additional concerns for slopes protected with vegetation include the possibility of fire following accidents, traffic hazards created by rabbits, and the costs of trimming and fertilizing vegatative cover.

The Arizona Department of Transportation was confrontedwith the problem of protecting hundreds of miles of freeway slopes from erosion in the arid parts of Arizona. In these areas annual rain fall measures less than 400 mm, but unfortunately, occurs frequently as torrential thunderstorms, (Climatology Laboratory 1988). In an attempt to mitigate slope erosion and realize economies, the State of Arizona became interested in the use of rock particles as an slope protection system. The authors were selected to develop a slope design program for rock slope protection.

A comprehensive research program was initiated to address such questions as how is erosion resistance affected by maximum particle size, shape of particles, specific gravity of particles, thickness of protection, slope angle, and protected soil properties. Early in the research program it became apparent that the durability of the rock particles was an issue that needed to assessed. Test techniques used by others were either too stressful in terms of mechanically applied forces or failed to model the rapid change in temperature and moisture state that these ground protecting particles would experience, ASTM Standard Test Method for Slake Durability of Shales and Similar Weak Rocks (D-4644-87).

Summer air temperatures, in the arid portions of Arizona can change from 50° C to 15° C within minutes when summer thunderstorms occur (Sellen et al. 1987). This 35 degree change in temperature is associated with a change in ground temperature of from 66° C to a temperature a few degrees above freezing when the storms frequently produce hail. The temperature change coupled with the transition from a dry to a saturated surface state produces stresses on exposed slope materials that increase the likelihood of physical as well as chemical weathering.

The on going research from the project showed that both maximum particle size and the percentage of rock retained on the 4.76 mm screen were important variables in slope transport resistance (Duffy and Hatzell 1989). A durability testing program, therefore, needed to detect changes in maximum particle size and the amount of minus 4.76 mm material being produced from the rock fragments. The State could not afford to place stronger material on the slope than was needed because of the large quantities of material needed. This testing program would have to determine the performance of the fragments when stressed in a way that specifically accounts for the effects of the summer storms.

Throughout much of central Arizona commercial sources of rock fragments are available as well as fluvial deposits containing a spectrum of igneous and metamorphic particles. As the research progressed several materials had favorable protective properties and were capable of being produced in sufficient quantities. Three sources of rock fragments or particles were selected for the durability testing program.

Three rock particles came from three local sources, Interstate 10 Granite (IG), State Route 360 Granite (SRG), and River Run from the Salt River (RR). These particles had the following descriptions:

IG materials: granite containing 55 % plagioclase, 30 % quartz, and 15 % biotite with other minerals.

SRG materials: granite-granodiorite containing 70 % quartz, 25 % feldspar, and 5 % other minerals.

RR materials (River Run, Salt River aggregates): fragments of chert, quartz pebbles, basalt, and a variety of sedimentary rocks.

The research showed that particle sizes greaterthan 4.7 mm and less than 50.8 mm provided effective protection from design storm intensities (Duffy and Hatzell 1989). Weathering of these coarse particles on the slope would increase the likelihood of smaller particles being produced to be transported down slope. Hydrolisis and physical weathering were expected to be the causes of the degradation of these particles. Because the particles would be exposed to water for total times measured in hours not days or years, the role of chemical weathering, other than hydrolisis, was expected to be minor. The stresses induced by rapid changes from dry and hot to wet and cold coupled with the diurnal temperature affects could exploit microfissures within the aggregates. The particles from the Salt River quarry were also thought to have a potential for smectite clay expansion in some particles.

To evaluate the cyclic stresses caused by summer thunderstorm activity, a test protocol was created that compressed years of thunderstorm activity into a three month testing program. The following test protocol was formulated:

1. Five aggregates that warranted consideration as slope protection were selected.

2. Each source was sampled to obtain representative samples weighing approximately 454 Kg. These samples were then sieved and that component larger than 4.7 mm and smaller than 6.4 mm was saved.

3. Each sample was split until six replicates weighing between 1 and 2 Kg were obtained for each of the five rock materials.

4. Three each of the replicates were used as control and three as treatment specimens. The three control replicates of each rock type were placed in a dust free environment and maintained at room temperature, which was approximately 32° C.

5. Once put into a sample container the particles were not removed until the testing was completed. During the weathering test cycle each treatment specimen was:

a) Placed in oven and maintained at constant temperature of 65.5 $^\circ$ C for 24 hours.

b) Removed from oven and allowed to cool to room temperature during a one hour interval.

c) Immersed in tap water for 23 hours at room temperature.

d) Drained and placed back in the oven for the next heating component of the test.

This environmental simulation cycle required 48 hours for completion. Interruptions in this cycle were kept to a minimum. Unavoidable interruptions were ocassionally made during the cool down part of the cycle. The samples were then maintained at room temperature until the soaking component could be restarted. The control samples were retained in their respective sample containers during the heating-cooling testing. The control sample temperature was maintained at room temperature while the heating-cooling portion of the testing was being conducted. The test temperature was restricted to 32 $^\circ$ C on the low end, instead of a colder value, because rain not hail is the most common product of the summer thunder storms.

The test duration was 50 cycles. At the completion of the last cycle both the control samples and the weathered treatment specimens were reweighed and the grain size distributions determined. The sieve nest used was: 50 mm. 25 mm, 19 mm, 12.5 mm, 4.76 mm, 2 mm, and 0.42 mm. Each sample was shaken for a 10 minute period.

The importance in establishing an adequate experimental control was recognized because of the 100 or more days required for the 50 cycles and sieve wear. Time was of concern because of the potential for wide ranges of humidity and temperature variations in Arizona over any 90 day period. This testing was started in the late spring and completed during the summer months. In an attempt to develop standardized testing conditions the same person was responsible for all sieving operations. Timer controlled shaking and same-person handling were expected to minimize experimental error. Control and treatment specimens from each group were sieved on an alternating basis to insure that sieve wear during the testing was not a factor in the testing procedure. Since the objective of the testing was not to develop precise grain size distributions but to detect differences between treated and control samples, sieve wear had to be negated.

To minimize abrasion the authors selected a single sieving after 50 cycles instead of sieving at intermediate cycle times. If the same samples had been sieved after 10, 20 and 30 cycles for example, abrasion may have shifted particle size distributions so that the test would not simulate the way the undisturbed particles would be stressed on the slopes. Sieving replicates would have increased the number of samples by 42 for each additional sample. A single sieving after a reasonable period of testing yielded the most useful and practical data. A test program with 50 cycles simulated approximately 5 to 10 years of thunder storms (Sellen et al. 1987). The obvious drawbacks to this cycled durability testing are the time required and the storage space necessary to protect both control and tested specimens.

The percent of weight passing each screen was calculated for each sample. The results of the grain size testing program for control and weathered samples is presented in tables 1 thru 5.

Table 1. The probabilities associated with comparison of control and treated samples of IG1 material

	Percent Passing		
Screen Opening mm	Control sample	Weathered Sample	t-Test Probability
50	100.0	100.0	1.00
25	100.0	100.0	1.00
19	99.8	100.0	0.37
12.5	99.1	98.1	0.48
4.76	3.5	4.0	0.01
2	0.3	0.7	0.03
0.42	0.2	0.4	0.07

Table 2. The probabilities associated with comparisons of control
and treated samples of IG2 material

Percent Passing

Screen Opening mm	Control sample	Weathered Sample	t-Test Probability
50	100.0	100.0	1.00
25	97.0	98.1	0.44
19	67.8	71.1	0.39
12.5	4.6	6.8	0.15
4.76	0.7	1.2	0.43
2	0.6	0.7	0.95
0.42	0.6	0.5	0.91

Table 3. The probabilities associated with comparisons of control
and treated samples of IG3 material

Percent Passing

Screen Opening mm	Control sample	Weathered Sample	t-Test probability
50	100.0	100.0	1.00
25	90.4	87.5	0.65
19	57.6	60.9	0.75
12.5	22.7	29.8	0.57
4.76	0.7	1.7	0.03
2	0.3	0.8	0.06
0.42	0.2	0.4	0.23

Table 4. The probabilities associated with comparisons of control
and treated samples of SRG1 material

Percent Passing

Screen Opening mm	Control sample	Weathered Sample	t-Test Probability
50	100.0	100.0	1.00
25	94.8	94.5	0.94
19	61.5	58.1	0.56
12.5	2.4	2.6	0.48
4.76	0.4	0.4	1.00
2	0.2	0.2	0.47
0.42	0.0	0.0	0.55

Table 5. The probabilities associated with comparisons of control
and treated samples of RR1 material

Screen Opening mm	Control sample	Weathered Sample	t-Test Probability
50	88.3	92.8	0.69
25	39.8	39.1	0.91
19	27.3	22.2	0.19
12.5	6.6	7.5	0.62
4.76	0.1	1.3	0.06
2	0.1	1.0	0.08
0.42	0.1	0.8	0.25

Percent Passing

Differences between the control sample and the weathered treatment samples for each material were evaluated with a t-test. The mean of the percent material by weight passing each screen size was calculated for the control and the weathered replicates of one material type. It was assumed that the variances for the control and weathered samples were the same. The t-test probabilities that the mean of the control for the IG1 material is not different from the mean of the corresponding weathered sample for each screen opening are given in table 1. For example, there is only a one percent probability that the mean of the IG1 control is the same as the mean of the weathered sample for the 4.76 mm size. In contrast, there is a 37 percent probability that the mean percent of IG1 material passing the 19 mm opening for the control is not different from the mean of the weathered sample.

The testing was used to screen out the IG2, IG3, and RR1 materials as possible sources of slope protection. The degradation of several of the larger particles of the RR1 samples was apparent during the testing program. These particles were sedimentary rock fragments that began to exfolliate after several cycles.

The effectiveness of this testing will be determined by the field performance of the aggregates selected for field application. The SRG1 and IG1 materials have been installed on freeway slopes. The performance of these materials has been monitored for a period of approximately four years. To date the examination of several test sections has failed to provide evidence of aggregate degradation.

The authors gratefully acknowledge the support of the Arizona Department of Transportation without who's support this research would not have been possible.

REFERENCES

Climatology Laboratory, Precipitation records for greater Phoenix", Arizona State University, Tempe, AZ 1988.

Duffy, D. M. and Hatzell, H. H., "Design of Slope Protection Systems and Maintenance Procedures to Minimize Erosion", Transportation Research Record 1189, TRB, National Research Council, Washington, D.C., 1989.

Sellen, W. B., Hill, R. H., and Ray, M. S., "Arizona Climate , 100 years", Department of Atmospheric Sciences, University of Arizona, Tucson, AZ, 1987.

David A. Lienhart[1]

THE MECHANISM OF FREEZE-THAW DETERIORATION OF ROCK IN THE GREAT LAKES REGION

REFERENCE: Lienhart, D. A., "The Mechanism of Freeze-Thaw Deterioration of Rock in the Great Lakes Region," Rock for Erosion Control, ASTM STP 1177, Charles H. McElroy and David A. Lienhart, Eds., American Society for Testing and Materials, Philadelphia, 1993.

ABSTRACT: The evaluation of the potential durability of rock materials in a cold climate is dependent upon a complete understanding of both the environmental conditions of exposure and the mechanics of the freeze-thaw deterioration. Information related to the three types of mechanical processes which result in deterioration of rock materials is presented. The four environmental factors and the three physical property factors which control the three mechanical processes are also discussed. It is shown that because of the variability of the controlling factors and their effect on the mechanical processes, it is extremely difficult to predict the potential durability of any rock used for erosion control. It is concluded that additional study is needed in development of a series of index tests which will provide a suitable/not suitable rating for use in a particular environment.

KEYWORDS: adsorptive suction, critical saturation, durability, freezing duration, freezing expansion, freezing rate, non-freezable water, porosity, volumetric expansion

---

The evaluation of the potential durability of rock

---

[1] U.S. Army Corps of Engineers, Geotechnical & HTRW Division, P.O. Box 1159, Cincinnati, Ohio 45201-1159

materials in a cold climate is dependent upon a complete
understanding of the interrelationship between and the
development of definitive information regarding:

a)   the environmental conditions of the expected site
of exposure, and
b) the mechanics of the physical and environmental
factors affecting freeze-thaw deterioration.

Two of the environmental factors affecting freeze-thaw
deterioration are freezing intensity and available moisture
(areas near coastlines, streambanks,etc. and high humidity
areas). The author (Lienhart 1988) attempted to define these
environmental factors for exposure within the contiguous
United States. Two additional environmental factors of
concern are freezing rate and freezing duration.

The mechanical or physical factors affecting freeze-
thaw deterioration are the tensile strength of the rock
(i.e., bursting strength of the rock material), critical
saturation, the total porosity and the pore-size
distribution (the capillarity of the rock material). These
factors, and their relationship to the mechanical processes
which affect freezing expansion, are discussed in more
detail following the description of the mechanical
processes.

PROCESSES AFFECTING FREEZING EXPANSION

Volumetric Expansion

This is the dominant process involved in "frost
shattering," "ice wedging" and "frost splitting" effects but
is not the only process. All of these effects come under the
classification of "congelifraction," a term generally
reserved for mechanical disintegration of rock due to the
pressure created when the water contained in pores and
fissures freezes (Bates and Jackson 1987). This process is
also dominant in "frost heave."

The 9% volume change which occurs when ice forms
produces a shattering force related to Young's Modulus in
tension. When the shattering force becomes greater than the
tensile strength of the rock, failure occurs. Due to the
fatigue caused by cycling between freezing and thawing,
failure may occur without the shattering force exceeding the
tensile strength of the rock (Mellor 1970). The pressures
involved with this process may best be illustrated by Table
1. This table shows the development of pressure with
decreasing temperature in a closed system of porosity under
fully saturated conditions ("closed system" is defined under
Adsorptive Suction, below).

TABLE 1--Pressures associated with volume increase
(after Powers 1965; Bowles 1982; and Ollier 1984).

TEMPERATURE, °C (°F)	PRESSURE, MPa
-1.1 (30)	14
-2.8 (27)	34
-5.6 (22)	69
-9.4 (15)	103
-12.5 (9.5)	138
-16.7 (2)	172
-21.7 (-7)	207

Adsorptive Suction

Although the adsorptive force or water migration theory
has been referenced in the literature as far back as 1930
(Black and Hardenberg 1991), it was not until 1983 that it
was actually observed (Fukuda 1983). The process involves
the movement of pore water toward the freezing front (the
boundary between the frozen and unfrozen portions of the
rock) by:
a) capillary action and,
b) suction due to negative pore pressure development
just ahead of the freezing front. This process may be
illustrated by Figures 1. and 2.

Figure 1. is an idealized graphical representation of a
rock outcrop. As freezing takes place in a partially
saturated rock outcrop, water from the smaller pores (pore
size is critical) will begin to migrate toward the inwardly-
moving "freezing front" (Figure 1b.) Ahead of the front,
negative pore pressures in the order of -10 kPa may develop
because of the freezing front's affinity for water. This
zone of negative pore pressures results in the suction of
additional water from the interior of the rock outcrop,
resulting in increased saturation at the freezing front
(Figure 1c.). Depending on the freezing duration and the
initial degree of saturation of the particle, the process of
volumetric expansion could become the dominant process and
result in development of fractures parallel to the face of
the outcrop. Because of the ability of the freezing front to
continue to move into the outcrop and to continue to draw
water from the interior of the rock mass, this particular
example is considered to be an "open system."

A rock particle, however, is considered to be a
"closed system" (Figure 2.). In a closed system, the same

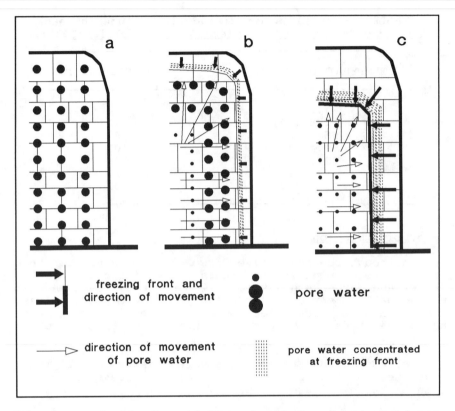

Figure 1--Idealized partially saturated rock outcrop (open system) in three stages of freezing; (a) prior to freezing; (b) after short freezing duration; and (c) after long freezing duration.

process takes place but with an important difference. In a partially saturated "closed system," as the freezing front continues its inward movement and saturation increases at the freezing front, a point will be reached where, with continued freezing duration, a zone of complete saturation will be concentrated at the rock's interior, the process of volumetric expansion will become the dominant process and the rock will burst. If sufficient moisture is not present at this time, the rock could absorb additional moisture during the next thawing cycle. This cycling process may continually add and subtract moisture until a point in time when the rock is critically saturated (i.e., enough moisture may be concentrated within the rock particle such that fracture development may occur).

The nature of the rock burst during freezing will depend on the degree of anisotropy of the rock's

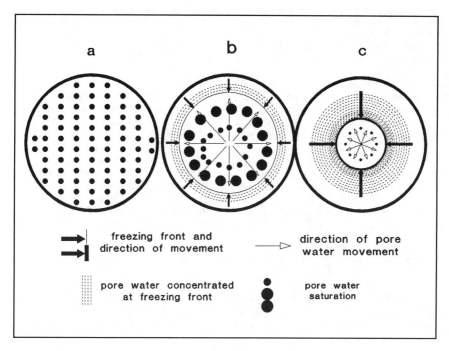

Figure 2--Idealized partially saturated rock particle (closed system) in three stages of freezing; (a) prior to freezing; (b) after short freezing duration; (c) after long freezing duration.

permeability. With isotropic permeability, the result is development of a "ball of ice" in the interior of the rock particle and radial fracturing from the interior of the rock. This has been observed and photographed in the field by Erickson (Lutton and Erickson 1992). With heterogeneous, bedded rocks with anisotropic permeability, the result is usually development of a lens of ice with the long axis of the lens oriented normal to bedding causing the rock to split normal to bedding.

<u>Expansion of the Unfrozen Water Phase</u>

The unfrozen water phase consists of the water occupying the smallest pores and finest capillaries. This water is typically considered to be connate water and can never be <u>totally</u> replaced if the rock particle is allowed to dry for a long period of time prior to being exposed once again to high moisture conditions (shoreline, high humidity areas, etc.); however, the water can be <u>partially</u> replaced through the process of adsorption. Connate water was referred to as "quarry sap" by old-time quarry workers and is the reason that freshly quarried rock was termed "green" and was set aside to "cure" (much the same as lumber is dried prior to use). As far back as the first century B.C. it was recognized by Vitruvius that stone needed to cure prior to use (Morgan 1960).

According to Dorsey (1940) and Winkler (1992), this water expands in volume by 0.6 percent as the temperature decreases from +4° C to -10° C . The pressures associated with this expansion are due to hydration of the capillary or adsorbed moisture. It is termed hydration water because of the ordered, almost crystalline state of this water and was first identified by Dunn and Hudec (1965). As with volumetric expansion, the theoretical hydraulic pressure associated with this unfrozen water is approximately 200 MPa. This implies that even if the rock particle does not contain sufficient moisture for volumetric expansion to take place, expansion of the non-freezable water can still result in disruption.

Additional implications of this particular process are as follows:

1. Rock quarried during temperatures of less than -2° C (critical freezing temperature according to Mellor, 1970) stand a greater chance of suffering some freezing damage or disintegration.

2. Rock quarried within a short period of time prior to the onset of temperatures of less than -2° C stand a chance of suffering some freezing damage or disintegration.

3. Some, but not all, rock needs to be "cured" prior to use.

The methodology required to determine which rock needs to be "cured" and which does not, was described by Lienhart and Stransky (1981) and the results are summarized by Table 2. For this test, freshly quarried pieces (averaging 25 to 30 kg in mass) of Berea Sandstone (Mississippian in age) were used. Specimen 1 was tested for 35 cycles (a cycle equals 12 hours freezing and 12 hours thawing) in a special freeze-thaw chamber (designed to test specimens of approximately 1,000 kilogram mass) to find the percent loss due to freeze-thaw on the fresh material. Specimen 2 was subjected to 20 cycles of wetting and drying (a cycle equals 6 hours wetting and 6 hours drying) prior to testing for 35 cycles in a freeze-thaw chamber. The wet-dry test simulates summer conditions of alternating rainfall and sunshine and was felt to best simulate accelerated "curing." Specimen 3 was subjected to 50 cycles of wetting and drying prior to freeze-thaw, and so on for the remaining specimens. Optimum "curing" for the Berea Sandstone appears to be about 100 cycles. The slight increase in loss for Specimen 8 at 200 cycles was attributed to the loss of a chip at the site of a pre-existing fracture. A complete description of the "freeze-thaw," the "wet-dry" and the "curing" test methodologies is given by Lienhart and Stransky (1981).

TABLE 2--Curing of freshly quarried Berea Sandstone (after Lienhart and Stransky 1981).

Specimen #	Curing Cycles, #	Loss During Freeze-Thaw, %
1	0	6.01
2	20	5.35
3	50	3.62
4	80	1.05
5	110	0.26
6	140	0.00
7	170	0.00
8	200	0.21

CRITICAL CONTROLLING FACTORS

As previously mentioned, there are several factors which have been found to be critical for any one or all of these three processes to take place. These critical factors may be divided into two categories: environmental (freezing intensity (temperature), freezing rate, freezing duration, and available moisture) and physical properties (critical saturation, total porosity and pore size, and Young's Modulus in tension). Because of the close interrelationships among these factors and the heterogeneity of rock as a natural material, the critical limits for many of these factors are variable from one rock type to another. Where critical limits exist, they will be given in the following discussion of each factor.

Freezing Intensity

Freezing intensity may be defined as the critical temperature for development of sufficient freezing strain to cause crack propagation. Walder and Hallet (1985) suggest that the range of -4° to -15° is most critical for crack propagation but their work was based on outcrop measurements. Mellor (1970) in his work with rock particles has shown that the freezing strain increases rapidly just below freezing but slows with decreasing temperatures. Mellor suggests a critical freezing intensity in the range of -2° C to -5° C. It is expected, therefore, that rock subjected to this critical temperature range will undergo the most intense crack propagation.

Freezing Rate

Because pore ice reacts plastically over a period of time, low freezing rate may result in low freezing strain and therefore, may result in less disruption to the rock particles. This factor may be useful in preventing disruption of rock particles in a vital portion of a structure if a way can be found to slow down the freezing rate at that particular point. There is no known critical value for this factor.

Freezing Duration and Available Moisture

These two factors are mutually interdependent. Walder and Hallet found that the duration of freezing is a critical factor in those circumstances where the rock is in intimate contact with a source of moisture (e.g., shoreline exposure). It has been shown that freezing strain will continue to increase for the duration of the freezing cycle as long as there is a source of available moisture, thereby creating a potential for significant damage in only one cycle. There are no critical limits applicable to this combined factor.

Critical Saturation, Total Porosity and Pore Size

The critical saturation factor is not only dependent on the available moisture factor but is intimately dependent on total porosity and pore size. The volumetric expansion and adsorptive suction processes of rock disintegration depend on a critical degree of saturation. Without sufficient moisture within the rock particle or available moisture to supply moisture through water migration to the freezing front, sufficient pressures for fracture propagation may not result. Dunn and Hudec (1965) found that a critical degree of saturation was 75 percent or greater for total porosities of 25 percent or less and for pore size distributions that would allow for an adsorption value of less than 45 percent at 100 percent relative humidity. For pore size distributions that allow for adsorption values greater than 45 percent, the criteria become more complex and are best summarized through the data presented in Table 3.

Young's Modulus in Tension

It is quite obvious that fracture development in frozen rock is directly related to both the freezing strain and the tensile strength of the rock in question. It follows then, that theoretically, Young's Modulus in tension is another factor that should provide some indication of the ability of a rock at greater than critical saturation to withstand

freezing at a fast rate for a long period of time at a
critical temperature. Early results of research in this area
indicate a possible correlation, but only with high strength
rock materials and the data are still somewhat "fuzzy."
Research in this area is continuing.

TABLE 3--Adsorption versus critical saturation and
total porosity (after Dunn and Hudec 1965).

ADSORPTION, %	CRITICAL SATURATION, %	TOTAL POROSITY, %
0 to 45	75 to 100	0 to 25
50	72 to 100	0 to 28
60	66 to 100	0 to 34
70	61 to 100	0 to 39
80	55 to 100	0 to 45
90	49 to 100	0 to 51
100	44 to 100	0 to 56

SUMMARY AND CONCLUSIONS

It has been shown that freeze-thaw durability can be
related to three processes that result in freezing strain:

a) volumetric expansion,
b) adsorptive suction, and
c) expansion of non-freezable or adsorbed water.

These processes are controlled by four environmental
factors;

a) freezing intensity,
b) freezing rate,
c) freezing duration,
d) available moisture;

and by three mechanical or physical properties;

a) critical saturation,
b) total porosity and pore size, and
c) Young's Modulus in tension.

With an understanding of the processes, the controlling
factors and their interrelationships, it is now possible to
gain an understanding of the actual deterioration mechanism.
In conjunction with this, the interpretation of certain
fracture patterns exhibited by various rock types under
freezing and thawing conditions are also possible.

Because of the heterogeneity of rock materials and the close interrelationships among the controlling factors, it is doubtful that any method of predicting the useful life of a particular rock material can ever be developed. It is recommended that instead, studies be directed toward the development of a series of index tests which will provide a "suitable/not suitable" rating for the rock material in a particular environment.

REFERENCES

Bates, R. L. and Jackson, J. A., Glossary of Geology, Third Edition, American Geological Institute, Alexandria, VA, 1987.

Black, P. B. and Hardenberg, M. J., Eds., Historical Perspectives in Frost Heave Research: The Early Works of S. Taber and G. Beskow, U.S. Army Cold Regions Research and Engineering Laboratory Special Report 91 -23, Hanover, NH, 1991, 169 p.

Bowles, J.E., Foundation Analysis and Design, Third Edition, McGraw-Hill Book Co., New York, 1982, 816 pp.

Dorsey, N.E., Properties of Ordinary Water Substrate, Reinhold Press, 1940.

Fukuda, M., "The Pore Water Pressure Profile in Porous Rocks During Freezing," Proceedings, Fourth International Conference on Permafrost, National Academy Press, Washington, D.C., 1983, pp. 322-327.

Lienhart, D.A. and Stransky, T.E., "Evaluation of Potential Sources of Riprap and Armor Stone - Methods and Considerations," Bulletin of the Association of Engineering Geologists, Vol. 18, No. 3, 1981, pp. 323-332

Lienhart, D. A., "The Geographic Distribution of Intensity and Frequency of Freeze-Thaw Cycles," Bulletin of the Association of Engineering Geologists, Vol. XXV, No. 4, 1988, pp 465-469.

Lutton, R.J. and Erickson, R.L., "Problems with Armor-Stone Quality on Lakes Michigan, Huron and Erie," Durability of Stone for Rubble Mound Breakwaters, American Society of Civil Engineers, New York, 1992, pp. 115-136.

Mellor, M., Phase Composition of Pore Water in Cold Rocks, U.S.Army Cold Regions Research and Engineering Laboratory Research Report 294, Hanover, NH, 1970, 61p.

Morgan, M.H. (translator), Vitruvius. The Ten Books of

Architecture, Dover Publications, Inc., New York, 1960.

Ollier, C., _Weathering_, Second Edition, Longman, Inc., New York, 1984, 270 pp.

Powers, T.C., _The Mechanism of Frost Action in Concrete_, National Sand and Gravel Association, Silver Spring, MD, 1965, 35 pp.

Walder, J. and Hallet, B., "A Theoretical Model of the Fracture of Rock During Freezing," _The Geological Society of America Bulletin_, Vol. 96, No. 3, 1985, pp 336-346.

Winkler, E. M., Professor Emeritus, University of Notre Dame, South Bend, IN, personal communication, 1992.

Ronald L. Erickson,[1]

EVALUATION OF LIMESTONE AND DOLOMITE ARMOR STONE DURABILITY FROM
OBSERVATIONS IN THE GREAT LAKES REGION

REFERENCE: Erickson, R.L., "Evaluation of Limestone and Dolomite
Durability From Observations in The Great Lakes Region," Rock For
Erosion Control, ASTM STP 1177, C. H. McElroy, and D. A. Lienhart, Eds.,
American Society for Testing and Materials, Philadelphia, 1993

Abstract: Laboratory testing is one of the tools one can use to predict
the long term durability of armor stone. Along with testing of the
stone, observations of the armor stones: rock type, bedding, quarrying
methods, methods of handling and placement, inclusions, voids or vugs,
saturation of the stone, clay content, overburden stresses, porosity of
the stone, time of year of production, previous production methods, and
experience of the quality control staff, should be used when making a
determination the suitability of an armor stone source. These
observations of armor stone, along with service comparisons of armor
stone produced under similar conditions can be used to improve the
selection of a source that produces armor stone that has the highest
long term durability.

Keywords: Durability, Armor Stone, Great Lakes, Performance, Freeze
Thaw, Pore Water, Deteriorate, Evaluate

INTRODUCTION

Armor stone is used to protect navigation projects, shoreline
structures, and other features subject to wave attack or erosion. Armor
stone can be as small as 50 kg, where wave energy is low, or as large as
30 metric ton, where the water is deep and wave energy can be very
large. Because of the lack of superior quality armor stone within close
proximity to the Detroit District Corps of Engineer's projects,
considerable efforts are spent evaluating the suitability of potential
stone sources. The geologist is often called on to determine if a
potential stone source is suitable for use on a particular project. The
common practice is to visit the source and select representative samples
for durability testing. After a period of time the laboratory test
results are evaluated and a conclusion is made on the acceptability of
stone from the quarry. Because of problems with stone breakage or
deterioration on past projects from sources that passed laboratory

---

[1] Geologist, Detroit District, U. S. Army Corps of Engineers, 477
Michigan Ave., Detroit, MI. 48226.

durability testing, additional efforts and evaluations are being made to determine if stone is suitable for use.

If the samples selected for lab testing were truly representative of the material delivered to the project, durability problems should not be encountered. Great efforts are made to assure that the stone submitted for testing is representative. However, many times the stone selected for testing is not representative of the product that is delivered to a project. If the stone tested is inferior to the delivered product, the only problem is that a suitable source could be deemed unacceptable. If the stone tested is better quality than the product delivered, many problems can develop (some short term and others long term).

Many properties of the stone determine the quality of armor stone as it finally rests in the structure. Some properties are naturally inherent in the stone, and others are the result of man's intervention.

Numerous tests have been developed to duplicate conditions an armor stone may encounter after placement. Lab testing for durability is determined based on accelerated weathering testing (freeze-thaw, wet-dry), absorption, adsorption, sulfate soundness testing, petrographic examination, and others. These tests give good guidance to the person making the determination of suitability; however it is felt that these laboratory tests do not duplicate the conditions that an armor stone encounters during it production, placement and in the environment. There are many observations and geologic evaluations that should be made to aid in making the determination of an armor-stone sources suitability.

DISCUSSION

The overall goal of a stone source evaluation is to assure the permanence in the structure and not have to perform maintenance for the design life of the project. If deterioration of the stone occurs, repairs can be very costly.

When a potential stone source is evaluated to determine the suitability of the source, the geologist should visit not only the quarry, but also visit the surrounding area, and examine places where the stone was previously used. The determination of suitability from a quarry or particular section of a quarry should not be a final determination, but should be continuously up-dated to monitor changes in the rock and or changes in the production methods that could affect the durability of the stone. The larger the size of the individual stone the more detailed the observations should be.

Following are examples of the observations that should be made.

Rock type:

The type of rock proposed is the first concern. Most igneous and metamorphic type rocks, that have been supplied to Detroit District projects have good durability. However, sedimentary rocks are more available in the Great Lakes area, and more quarries have been established in this types of rock. Most quarries in the area have some stone that could be suitable, but the quantity may be limited.

Bedding and Spacing of Beds:

The natural bedding of the stone and the spacing of the bedding controls the size of stone that can be quarried. Sometimes the bedding is quite obvious and easily seen, while other times the bedding is hard to distinguish. Often in sedimentary rock bedding may be continuous, and the stone separates along the bedding. Jointing and faulting of the source rock also control the size of stone that can be produced. The bedding in a source often limits the quarry to producing armor stone only from particular layers in the quarry.

Quarrying Methods:

An important observation that determines the quality and durability of stone is the method in which the stone is quarried. Many limestone quarries are developed primarily for the production of small size stone and aggregates. Often the stone is blasted from a high face with large diameter holes and loaded with explosives from top to bottom. The resultant stone from this type of blasting often contains blast fractures that breakdown upon successive freezing and thawing. The most desirable quarrying operation would have a low face height with closely spaced, small diameter drill holes, which are lightly loaded or are deck loaded with delays to reduce the effects of the blast energy. A blast from this type of operation would appear to not break the stone but only separate the stone from the face, along existing bedding, fractures or joints.

Method of Handling and Placement:

Another observation that should be made for the production of durable stone is the method by which the stone is handled, transported to the job site, and placed in the structure. Because the quarry source is often many miles from the job, the stone is transported in many fashions. For a recent project the stone was conveyed from the quarry by truck, to a rail site, dumped on the ground, loaded onto rail cars, transported to a dock, and dumped again into stockpiles where it was then loaded onto barges for transport to the construction site. At the construction site the stone was then handled a final time when it was placed into the structure. Each one of these steps potentially damages the stone and decreases its durability. Ideally the stone should be quarried and transported without having to change modes of transit. When stone is dumped from trucks, it should be carefully dumped to avoid excessive fall heights, which can cause durability problems. At the job site the stones should be carefully placed to avoid any damage. To prevent damage or breakage of the stone, our specifications limit the height a stone can be dropped to one foot.

Inclusions (Detrital Material, Chert, Etc.):

Often the overall quality of the sedimentary rock is quite good, but inclusions in the stone make limestone or dolomite poor quality or unacceptable. Chert is the primary inclusion that severely affects the durability of the stone. Because the chert has a high absorption rate, the chert expands and causes cracks in the stone. Other inclusions include fossil beds, mud, and siltstone (if they are continuous or form weak planes in the stone).

Voids or Vugs:

Vugs are holes in the stones and can be formed in many ways. Often in sedimentary rocks the holes are the result of groundwater solution. Individual voids or vugs are generally not a problem with the durability of the stone. In fact, it has been our opinion that small vugs evenly distributed throughout stone can be beneficial as they act similar to air entrainment in concrete and allow for expansion of freezing water into the voids. The problem with voids or vugs in stone are when they become too large or they are segregated along a particular plane. They then act as a plane of weakness and cause the stone to be less durable.

Saturation of Stone - Groundwater Table Elevation:

Another problem that one should be aware of is the saturation of the stone. One should look at the quarry faces to see if there are signs of water seeping into the quarry. Often the elevation of the water table is higher than the quarry floor. The surface of the groundwater table can often be seen in the face as a distinct layer below which the stone is stained. A good time of year to see the saturation of the stone is in the winter. Because the water is not frozen behind the face, when it seeps out along the face, it will then form large icicles down the face. The effects of water in stone has been a concern for a long time. If a source is found to have significant amount of water flowing into the quarry, the source of the water should be evaluated. If the water is only moving along faults or fractures in the rock, it is less of a potential problem then if the water is seeping through the pores of the rock. When the stone is saturated with water and subjected to freezing, the build up of pressure from the expansion of the freezing water often cracks the stone. Because of this concern, specifications may require stone to "cure" to allow the saturated stone to lose pore water prior to being subjected to freezing conditions.

Clay Content - Stylolites:

The durability of armor stone in the field is quite often affected by thin layers called stylolites or suture joints. These thin layers are believed to be the result of seasonal fluctuations during the deposition of the sedimentary layers. Similar to stylolites, thin layers of silt or clay often make weak bedding planes in the rock. Stylolites are often continuous over a large area, and can control breakage of the stone. The occurrence of these layers needs to be thoroughly examined to determine the detrimental effects. The acceptance of a quarry for production of armor stone may be dependant upon the area and frequency of occurrence of stylolites.

Overburden Stresses - Stress Relief:

The rock layers in the Great Lakes area have been subjected to stresses and consolidation due to the weight of the glacial ices that moved across the area since the rock layers were deposited. This consolidation and subsequent rebounding of the bedrock has caused fracturing and a build up of stresses in the rock. When quarrying the rock, this built up stress is released and can cause fracturing of the rock. To determine if the stone from a particular quarry is subject to stress relief is difficult. One indication is if the quarry floor experiences heave or rebound after blasting. Also one can often see the effects of the stress by looking at the rock immediately below the bedrock surface. Often the rock has many layers which are very thin near the surface, and increase in thickness with depth.

Porosity of the Stone:

The durability of the stone is partially dependent upon the porosity of the stone. Sedimentary rocks often have considerable porosity, and if the stone is saturated, the effects of freezing can be quite damaging. It is difficult to observe the porosity of the stone in the field if the pores are small; however one can see the effects of freezing on stone that is saturated. Stone that is frozen when saturated often cracks, with cracks radiating from the center of the stone.

Time of Year of Production:

The time of year that a stone is produced can also affect the durability

of the stone. The stone from several limestone quarries that was
quarried during freezing temperatures was severely damaged when the
stone was frozen. When quarrying in the winter months, the stone back
from the face is not frozen and the temperature is approximately 50
degrees. When it is blasted the unfrozen groundwater in the stone, is
quickly frozen, when exposed to the elements. When water freezes it
expands approximately nine percent. Since the freezing of the stone
begins on the exterior of the stone, the expansion of the pore water
seals the pores of the stone. As the freezing progresses toward the
center of the stone the expansion of the water does not allow the
expansion of the pore water in the interior of the stone. Discussions
with people at the Corps of Engineers, Waterways Experiment Station
indicated that research on the freezing of concrete has shown that pressures up to 5,000 psi can build up in the interior of concrete. This
pressure can easily crack a stone, particularly if the stone already has
blast induced fractures. Because of this concern, our specifications
limit the time of the year that stone can be quarried.

Previous Production Methods:

Another potential question that can be asked is how the stone, that was
previously shot from the face of the quarry, was produced. Several
times, we have found that the method of production of the previous stone
effects the quality of the armor stone. This is particularly important
when selecting a stone for lab testing, since the stone is probably
blasted for a smaller gradation. Since most quarries do not commonly
produce armor stone, their normal blasting operation uses high-blast
energies. This high-blast energy not only blasts the rock away from the
face, but the blast can cause fractures back into the new face. This
heavy blast affects stone subsequently produced with low-blast energy.
When the quarry changes to production of armor stone, the face that they
are trying to produce stone from, has been fractured by the previous
production. To avoid this problem it is suggested that the quarry blast
the face with a low energy blast, and use the stone from this blast for
other than armor stone.

Experience of Quarry Operator and Quality Control Staff:

One of the main items that can control the durability of the stone
produced from a quarry is the experience of the quarry operators staff
and the quality control staff. Most quarries, in our area, have very experienced production staffs. However, sometimes the experience is
limited to the production of small sized stone. When asked to produce
large armor stone, the quarry may try to produce the large stone using
methods derived from production of small stone. Also, quite often the
quarry's experience and equipment is geared for drilling large diameter
holes, and large blasts. The quality control staff oftentimes is
experienced in the problems associated with small stone, and when they
produce large stone they may overlook cracking, bedding or handling
methods that lead to poor quality stone.

Service Record of the Stone:

The last item that should be carefully observed is the service record of
the stone. The geologist should find places where armor stone from the
potential quarry has been previously used. This stone should be from the
same layers, produced in the same method, and during the same period of
the year. Evaluation of this existing stone can often reveal problems
that could not be seen in the stone at the time of production. To track
the usage of stone, the Detroit District has recorded the stone
production methods, sizes, and problems encountered from each of the
quarries it has used. These records are becoming very useful in

determining the quality and durability of stone produced from a particular quarry.

CONCLUSIONS

The use of laboratory testing is one of the tools that should be used to determine the durability of armor stone. The observations listed above should also be used. Often the lab tests are run on samples that may not be representative of all of the stone that is produced from a particular quarry. The use of lab testing along with field observations and performance comparisons of similar produced stone should better insure the production of good quality, durable armor stone. Lab testing has guidelines as to when a stone is acceptable or not. Field observations are somewhat more subjective, and therefore should be made by an experienced geologist.

G. S. Wong[1] and R. J. Lutton[2]

PETROGRAPHIC EXAMINATION OF LARGE STONE FOR DURABILITY

REFERENCE: Wong, G. S., and Lutton, R. J., **"Petrographic Examination of Large Stone for Durability,"** Rock for Erosion Control, ASTM STP 1177, C. H. McElroy, and D. A. Lienhart, Eds., American Society for Testing and Materials, Philadelphia, 1993.

ABSTRACT: The ASTM Standard Guide for Petrographic Examination of Aggregate for Concrete (C 295) includes provisions appropriate for the examination of large stone in Section 11.1 on Ledge Rock. That standard outlines procedures to be used. In this paper information obtained from petrographic examination is shown to be very helpful in estimating rock quality particularly among sedimentary rock types. Composition and microstructure characteristic of satisfactory rock and of unsatisfactory rock are identified. Evaluations based on experience continue to be important.

KEYWORDS: durability, freezing-thawing, petrography, riprap, rock, stone

INTRODUCTION

Several tests and examinations are conducted by the Corps of Engineers (CE) to determine the quality of stone for use in slope protection. One such test is the freezing-and-thawing test originally developed by the Waterways Experiment Station (WES) for the Omaha District (Mather and Mather 1962). Other tests, mostly adapted from those used in testing aggregate, include specific gravity, absorption, sulfate soundness, and abrasion resistance.

Petrographic examination usually is made an integral part of the laboratory testing and characterization; however, under some circumstances it may be conducted alone. Testing such as for resistance to freezing and thawing or wetting and drying or both generally requires a month or more to complete. The engineer often needs some indication of the quality of the stone prior to the completion of such testing.

STEPS IN EVALUATION

Corps of Engineers practice in obtaining suitable riprap stone is outlined in the CE Guide Specification on Stone Protection for Slopes and Channels (Corps of Engineers 1958).[3] That guide specification states that "Stone for riprap and derrick stone shall be durable and of a suitable quality to insure permanence in the structure and in the climate in which it is to be used. It shall be free from cracks, seams,

---

[1]Petrographer, Structures Laboratory, US Army Engineer Waterways Experiment Station, Vicksburg, MS 39180-6199.

[2]Geologist, Geotechnical Laboratory, US Army Engineer Waterways Experiment Station, Vicksburg, MS 39180-6199.

[3]Other guidance is reviewed in another paper in this volume by Lutton and Wong.

and other defects that would tend to increase unduly its deterioration from natural causes."

The steps in evaluation of available sources of stone for a particular project frequently go as follows (Lutton 1990):

1. Preliminary examination by geologists, petrographers, and materials engineers of locally and economically available sources.
2. Consideration of performance of the material from these sources in service, in road cuts, outcrops, and stockpiles or, as necessary, of stone of similar lithology. At this stage some sources may be eliminated.
3. Detailed examination of the sources not eliminated during step 2 with the selection of samples for laboratory study.
4. Petrographic examination and physical tests in the laboratory. The physical tests should include specific gravity and absorption plus such performance-simulating tests as may be deemed appropriate.
5. Judgments can be made based on the results of steps 1 through 4 to list the source for further consideration or to reject it.

TECHNIQUES AND CRITERIA

When a petrographic examination is performed, guidance for the examination can be obtained from "Standard Guide for Petrographic Examination of Aggregates for Concrete" (C 295) in Section 11.1 on Ledge Rock. Additional guidance for nomenclature is presented in "Standard Descriptive Nomenclature for Constituents of Natural Mineral Aggregates" (ASTM C 294).

Ideally, at least three large pieces of rock should be taken to represent each working face. Where there are separate working ledges, three large pieces should be taken from each ledge; and where there are multiple rock types, three large pieces of each rock type are desired. The size of each piece should be similar to the size of the pieces that will be produced for the intended final use.

Where multiple pieces are received, first put the pieces in categories to minimize testing. Select and mark pieces for testing based on the categories and the requirements of the testing program. Specimens for testing for resistance to wetting and drying and to freezing and thawing should be marked for cutting normal to bedding.

Prior to any processing the as-received samples are examined and visible characteristics are recorded. Observations should include the following:

- size (length, width, thickness)

- particle shape, characterized as: cubic, blocky, spherical, ellipsoidal, pyramidal, flat

- bedding thickness

- joint spacing

- appearance (color, weathered, fresh, joints, fresh fractures, seams, cavities and pores, stylolites)

- composition (nodules, lenses, material difference, bedding planes)

Dense massive homogeneous rock free of cracks, joints, and seams would usually be expected to be durable. These features can often be identified in the initial examination of the samples. Pieces may be weathered or coated so as to obscure these physical details and may require supplemental examination of freshly fractured surfaces. A sawed surface offers enhanced details that are not as easily observed in the material as received.

The constituents must be chemically and physically stable for expected weathering conditions. Stone containing water-soluble minerals such as anhydrite should be considered with special care or avoided. A general rounded appearance of crushed stone that has been stockpiled may give an indication of solubility or physical weakness.

Acid etching is useful in the examination of carbonate rocks such as limestone and dolomite. Etching with dilute HCl reveals noncarbonate minerals as the carbonates dissolve and the noncarbonates remain in relief. Structural features tend to be emphasized on etched surfaces. The WES laboratory uses a solution of 1 part concentrated HCl to 4 parts water for etching. Specimens may be treated with ferric chloride and sodium sulfide solutions to determine the calcite content since only calcite is stained.

Some types of noncarbonate minerals, e.g., clays that would be regarded as detrimental, may be tolerable where the amount is small and the minerals are well disseminated through the groundmass.

The presence of seams, veins, cracks, bedding planes, nodules, and other discontinuities should be regarded with suspicion. These potential weaknesses should be investigated for stability in the environment in which the rock will be exposed.

Temperature variation during exposure has been suspected (Lutton and Erickson 1992) of being a cause for disintegration of stone. Pieces of limestone containing chert nodules or stringers, although seemingly competent, apparently degrade under stresses caused by changing temperatures because of differences in coefficients of thermal expansion between chert and carbonate minerals.

Seams containing clay minerals are often sources of weakness. If the seams are thick, regardless of the type of clay, there is a likelihood that the rock will separate along the seams. Expansive clays, especially if found in continuous seams, will often cause opening during wetting. These clays can be identified using X-ray diffraction in a two-step procedure. A clay sample is examined by X-ray diffraction dry and then again following treatment with glycerol (Fig. 1). Movement of either a 1.4-nm peak or a 1.2-nm peak suggests the presence of expansive clay. There are other peaks that may also indicate potential problems.

Minerals can often be identified in oil immersion mounts. In recent years X-ray diffraction has provided a more direct approach in the identification of minerals and has freed the petrographer for visual analysis of physical features.

Examination using a stereoscopic microscope reveals details of the microstructure. Pores, grain interlock, and matrix structure can give indications of the durability of the stone. Rock with large open pores may seem susceptible to frost damage but actually is often durable. The open pores drain faster than the water freezes and no internal pressure develops. Where interconnected pores are of capillary size, restricted water movement can lead to deleterious expansion during freezing of critically saturated stone.

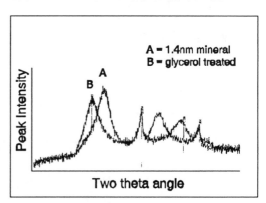

Fig. 1--X-ray diffractogram with peak shift in expansive clay.

To gain some insight on the porosity, a sample can be impregnated using an epoxy containing a fluorescent dye. Thin-sections of samples treated in this manner and examined using a UV light source will show the capillary pores.

Before Test

After Test

Fig. 2--Effects of freezing-and-thawing on quarry sample slabs. Rock consists of fine-grained calcitic dolomite containing thin layers rich in quartz, clay-mica, kaolinite, feldspar, and hematite.

After Test

Fig. 3--Effects of freezing-and-thawing on quarry sample slabs. The rock is composed of lightly weathered silica-bonded sandstone containing scattered voids and kaolinite disseminated through the rock in the interstitial spaces. The spalling of corners suggests that the porosity and interstitial clay are detrimental.

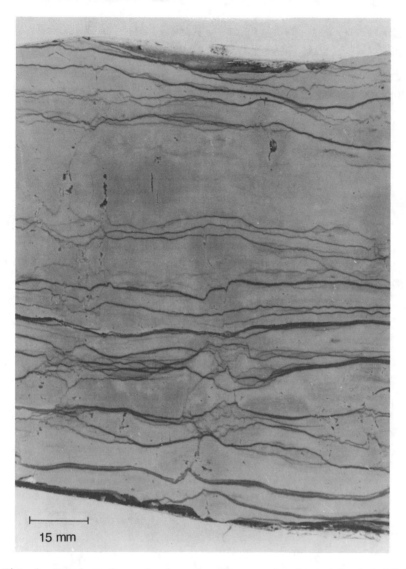

15 mm

Fig. 4--Sawed surface showing closely spaced, clay-rich stylolitic partings in limestone. Vulnerability of stone to breakdown is roughly proportional to the abundance of such partings.

Fig. 5--Brushing sawed surface of dolomitic limestone removed soft shaly pockets and accentuated angular pits of potential concern for durability.

20 mm

Fig. 6--Sawed surface showing finely crystalline, very porous
argillaceous, dolomitic limestone (light gray), containing lenses of
cross-bedded sandy dolomitic limestone (dark gray). Sawed surface
allows for better evaluation of small-scale bedding and porosity as
potentially detrimental features.

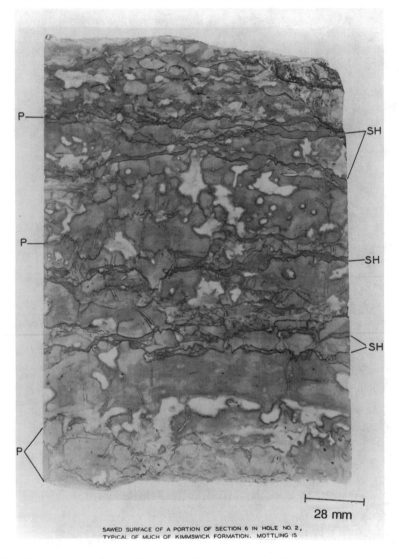

SAWED SURFACE OF A PORTION OF SECTION 6 IN HOLE NO. 2,
TYPICAL OF MUCH OF KIMMSWICK FORMATION. MOTTLING IS

Fig. 7--Sawed surface of complexly structured limestone reveals natural
solution and deposition of carbonates. Shale seams (SH) and stylolite
partings (P) make this rock of doubtful durability for use as riprap.

5.5 mm

Fig. 8--Ground surface of dense limestone with open crack following a
stylolite. The crack constitutes a well-defined flaw in the stone.

Fig. 9--Ground surface in fine-grained carbonate rock reveals a set of natural open cracks. Such cracks constitute flaws which may extend even more during blasting in the quarry.

⊢————————⊣1.7 mm

Fig. 10--Sawed surface showing beds of dolomite (dark gray) alternating
with beds of porous chert and clay (light gray). Rocks containing
appreciable clay or porous chert are ordinarily unacceptable for riprap.

Fig. 11--Etching of sawed surface with warm HCl accentuates the seams of porous chert (white) in finely crystalline dolomitic limestone.

Fig. 12a--Sawed surface after treatment with ferric chloride and sodium sulfide solutions. Finely crystalline calcite (stained black) has been replaced by stringers of well-crystallized dolomite (unstained, white).

Fig. 12b--Photomicrograph of a small portion of 12a showing the well-crystallized dolomite in the stringers. Throughout the specimen where dolomite and calcite occur together, the dolomite is in the form of well-developed crystals.

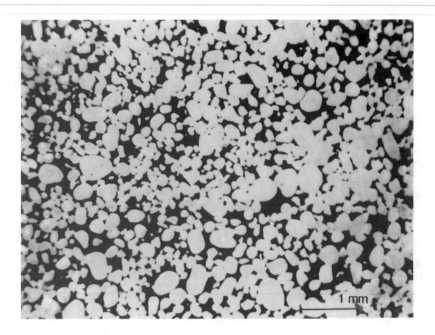

Fig. 13--Sawed surface of sandstone after treatment with ferric chloride
and sodium sulfide solutions. Rounded grains of quartz (white,
unstained) are contained in a matrix of calcite (stained black).
Reflected light. Low porosity is potentially favorable for durability.

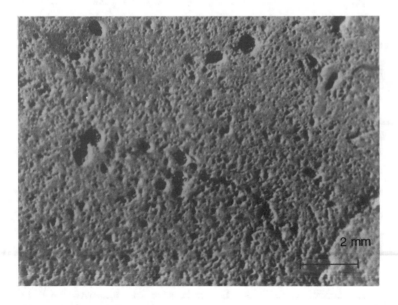

Fig. 14--Ground surface reveals scattered voids in limestone. Oblique
reflected light. Smaller capillary pores are potentially unfavorable
for durability.

PHOTOGRAPHS

Figures 2-14 have been included as stand-alone illustrations of features and procedures that may prove important in riprap examination and evaluation. Together the figures also illustrate the importance of documenting petrographic findings with photographs. As shown above, petrographic methods are particularly well suited to evaluating sedimentary rocks where subtle differences can dramatically affect performance as riprap; hence the emphasis on sedimentary rocks among the illustrated examples.

SUMMARY

Petrographic examination is a powerful tool for determining the quality of stone being considered for riprap. Petrography can stand alone when circumstances permit or require. Otherwise, the petrographic examination serves to link all of the information generated in the laboratory. Eventually, the petrographic examination is integrated with historical and field information as well as laboratory test results in a broad materials study.

ACKNOWLEDGEMENT

The techniques presented herein were compiled for presentation under the Civil Works Guidance Update Program of the United States Army Corps of Engineers. Permission was granted by the Chief of Engineers to publish this information.

REFERENCES

Corps of Engineers, 1958, "Stone Protection for Slopes and Channels," Civil Works Guide Specification CE 1308, Washington, D.C., 19 pp.

Lutton, R. J., 1990, "Material Characteristics of Large Stone in American Construction Practice." Vol. 286, Transactions of Society of Mining, Metallurgy, and Exploration, pp. 1850-1855.

Lutton, R. J., and Erickson, R. L., 1992, "Problems with Armor-Stone Quality on Lakes Michigan, Huron, and Erie," in Durability of Stone for Rubble Mound Breakwaters, American Society of Civil Engineers, pp. 115-136.

Mather, K., and Mather, B., 1962, Evaluation of Stone for Protection Work, Miscellaneous Paper No. 6-480, US Army Engineer Waterways Experiment Station, Vicksburg, Miss., 15 pp.

Specification Conformance Testing

Herman Schauberger,[1] Colin O.D. Arrand,[2] and Jeffrey D. Major[1]

PRODUCTION OF EROSION CONTROL STONE

REFERENCE: Schauberger, H., Arrand, C. O. D., and Major, J. D., **"Production of Erosion Control Stone,"** Rock for Erosion Control, ASTM STP 1177, Charles H. McElroy and David A. Lienhart, Eds., American Society for Testing and Materials, Philadelphia, 1993.

ABSTRACT: The methods used in the production of erosion control stone depend on the size and quantity of the product required. Discussed in this paper are the blasting and various sizing methods used with a detailed description of a high volume production plant. The quality control that needs to be performed both before and after production is reviewed and suggestions made for reducing field problems.

KEYWORDS: erosion control, stone size, durability, blasting, grizzly bars, perforated plates, gradation

INTRODUCTION

   The production of erosion control stone using only manual labor for the selection, sizing, and loading has not been considered feasible since the early 20th century. Methods, utilizing shovels/loaders for selecting, sizing,

[1]Director of Internal Services and Geologist, respectively, Reed Crushed Stone Co., Inc., a subsidiary of Vulcan Materials Company, P. O. Box 35, Gilbertsville, KY 42240.

[2]Construction Materials Engineer, Research and Development Laboratory, Vulcan Materials Company, P. O. Box 530187, Birmingham, AL 35253-0187.

and loading the stone are still in use today. There are
limitations to these techniques, however, in that selection
of the proper size stone depends on the careful examination
of each stone to maintain specific grading or weight requirements.

The most advanced method used today begins by customiz-
ing the blast based on the specific grading of large stone
required. This is accomplished using various combinations
of drill hole patterns and powder factors. The stone is
then loaded and hauled with modern equipment to a scalping
facility. The scalping facility may consist of one or more
grizzly bar sections (Figure 1), situated either before or
after a primary crusher, that remove desired sizes before
further crushing and screening operations. The facility
may employ no crushing and consists only of a series of
large screens or grizzly sections which classify the stone.
Any material not selected for erosion control can be used as
plant feed for production of other stone products.

FIG. 1--Typical grizzly bar scalping section
in a plant. Bar spacing is 20 cm
(8 in.).

QUARRY SELECTION

The first step in the production of erosion control
stone is the selection and approval of the quarry or quarry
face to be mined. The stone formation being quarried must
be such that it is capable of producing stone of the desired
size and with the required durability characteristics. For

example, in order to produce durable breakwater stone of large dimensions from a limestone formation, a tightly ce- mented, uniform, hard strata is necessary. Shale, chert seams and styolites are avoided. For igneous lithology, if the in-place rock structure is unfractured, the size of stone is primarily dependent on the blasting technique. For metamorphic rocks there may be laminations in the rock due to micaceous layers. These may provide weak points which may limit stone size in quarrying or prove nondura- ble.

Not all of a quarry face may be suitable for the pro- duction of erosion control stone. Selected use of one lift or section of a lift may result in increased production costs. In order to assess the potential of a quarry face for the production of erosion control stone, an initial geological evaluation should be performed by a professional geologist. If, in his/her opinion, the face or section of a face will likely produce stone of the desired size and durability, then durability testing of the areas under con- sideration should be initiated.

The required durability of erosion control stone is usually determined by specifications. These can vary con- siderably but usually bear a relationship to its end use. For example, for breakwater stone used in the Great Lakes region the Corps of Engineers' stringent specifications not only ensure long-lasting structures but also minimize cost because of the logistics required to replace failed stone in an existing structure. In order to determine the dura- bility of stone from a particular quarry the Corps of Engi- neers will select samples from the quarry face from which they can cut 76 cm (30 in.) square samples which cross any bedding planes in the rock. These samples are then tested for durability by subjecting them to wet-dry, freeze-thaw and sulfate soundness testing.

Most laboratories have the capability to subject normal size construction aggregates to the above-mentioned tests but would find it impossible to follow Corps of Engineers procedures using large specimens. In order to make it more practical for all laboratories to test the durability of erosion control stone, ASTM has proposed tests using speci- mens that are 14 cm (5.5 in) square. Once this durability testing is complete, the quarry or quarry face can be ap- proved or disapproved for production.

PRODUCTION

Blasting

A wide variety of drill hole patterns and explosive factors are utilized in the quarrying industry to produce erosion control stone because of the diversity of rock types

and formations being mined. The density of the drill hole
pattern and the powder factor used in a specific formation
is determined by the geological conditions present and the
maximum fragment size necessary to meet the grading requirements of the end product.

In the past, the choice of hole density and powder
factor was determined only by experimentation in the field.
The blasting operation may now be optimized and cost-effec-
tively accomplished by utilizing computer programs to actually simulate a shot prior to its occurrence. The program
can estimate various levels of fragmentation and gradings
which result from use of various combinations of drill hole
patterns and powder factors. Although the computer program
has the ability to simulate the blasting sequence, the spacing of the drill pattern and the powder factor will still
vary until the desired quality of fragmentation is obtained.

Sizing

Production techniques will vary among producers of
erosion control stone based on the number of different
gradings and the quantity of this type of stone needed to
meet market demands. Production methods include:
a) manually measuring each stone needed to produce a
particular grading, b) utilizing loaders/shovels for
obtaining the desired gradings, c) using scalping
facilities where crushing is involved, and d) using modern
scalping facilities which mechanically separate the stone
based on its individual size when mechanical crushing is not
involved.

Production of erosion control stone using methods where
each stone is individually measured in size to determine if
it is within specification limits is very labor-intensive
and costly. It requires a considerable amount of time to
obtain a minimal quantity of the desired product. Although
used, this method would not be considered a practical production method when a large volume of a particular grading
is required within a short period of time.

Erosion control stone may also be separated or graded
by the use of loaders/shovels for "picking" individual
stones which meet a particular grading criteria. Larger
volumes can be produced much faster than in the previous
method, but maintaining accuracy within a specific grading
is limited to the skill of the loader/shovel operator.
Some operations have adapted the loader/shovel buckets to
include a section of grizzly bars in the bucket bottoms
(Figure 2). This allows the larger stone to be retained in
the bucket while allowing the small material to fall through
and remain in the stockpile. These buckets are usually

custom-built in order to meet the needs of a particular
project.

FIG. 2--Front-end loader with special grizzly
bar bucket

The third method of producing erosion control stone is
through the use of scalping facilities which consist of one
or more grizzly sections situated either before or after a
primary crusher. By allowing the stone to be processed
through a primary crusher, the maximum size will be no
larger than the maximum crusher opening. The minimum size
will depend on the grizzly bar spacing and/or hole size in
the perforated plates where the stone is separated. The
nearer these two dimensions are together, the narrower the
range of stone size would be included in a specific grading.
This method of production is as accurate as either of the
previous ones, allows large volumes of stone to be processed, and allows a constant grading to be maintained.

The fourth method of production is the use of an independently operating plant which employs no crushing. It
consists of a series of large screens and/or grizzly sections which are utilized for classifying and separating the
stone based on its specific dimensions (Figure 3). Figure 4 is a flow schematic of this type of facility in use at
the Reed Stone quarry. The stone goes into a receiving
hopper where it is transported by an apron feeder onto the
first grizzly feeder. This feeder has a 61 cm (24 in.)
grizzly bar spacing which directs all material larger than

61 cm (24 in.) across the top of the grizzly feeder and down
a chute into a stockpile. All material less than 61 cm
(24 in.) falls through onto the second grizzly feeder with a
38 cm (15 in.) grizzly bar spacing. This directs all material within the 38 cm to 61 cm.(15 in. to 24 in.) size
range over the grizzly and onto a conveyor and into a stockpile. All material smaller than 38 cm (15 in.) falls onto
a chute which directs the material onto the scalping screen.
This particular screen utilizes perforated plates with 20
cm (8 in.) diameter round holes. All stone within the 20 cm
to 38 cm (8 in. to 15 in.) range is directed over the
perforated plate and onto a stockpile conveyor. All
material smaller than 20 cm (8 in.) is used as plant feed
material to produce construction aggregate. This type of
facility is designed for mass production of erosion control
stone and maintains an extremely high rate of accuracy and
consistency within the various gradings.

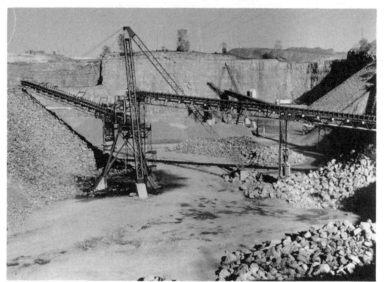

FIG. 3--Large capacity erosion control stone
production facility.

Quality Control

Quality control of erosion control stone is mainly concerned with the grading of the product, the durability of
the product having been verified by pre-production testing.
However, when armor stone is being produced it is examined
for cracks and usually a holding period is required to allow
the development of stress relief cracks.

The grading of a product may be specified and/or tested
in terms of weight or size, these two measurements being

readily converted using the specific gravity of the material. Whether erosion control stone is tested by weighing or

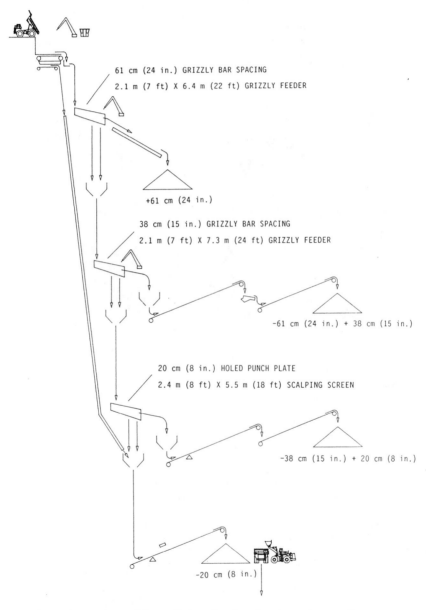

61 cm (24 in.) GRIZZLY BAR SPACING

2.1 m (7 ft) X 6.4 m (22 ft) GRIZZLY FEEDER

+61 cm (24 in.)

38 cm (15 in.) GRIZZLY BAR SPACING

2.1 m (7 ft) X 7.3 m (24 ft) GRIZZLY FEEDER

-61 cm (24 in.) + 38 cm (15 in.)

20 cm (8 in.) HOLED PUNCH PLATE

2.4 m (8 ft) X 5.5 m (18 ft) SCALPING SCREEN

-38 cm (15 in.) + 20 cm (8 in.)

-20 cm (8 in.)

FIG. 4--Schematic of typical plant set up for
producing erosion control stone.

measuring its dimensions is often decided by the size of
the product. It is often easier to lay out large stone on
the floor of the quarry and measure each piece (Figure 5)
than to load each piece onto a harness and portable scale
attached to a crane to obtain its weight. One advantage of
measuring the product's dimensions is that this procedure
also allows the checking of shape of the stone since some
specifications limit the amount of flatness and elongation
in each rock.

FIG. 5--Armor stone laid on quarry floor for
quality control inspection.

Another technique for determining stone size has been
to use front-end loaders with the truck-weighing scales at
the plant. Sometimes pre-weighed stones are used to visually compare production material in order to sort out different sizes. It is obvious that as the size of product being
tested decreases, the testing options increase because of
the greater ease of handling smaller pieces.

A major cause for problems occurring on erosion control
projects is the point at which the quality control of the
product be performed. From the producer's point of view, it
is preferable that quality control is performed at the plant
to ensure production is within specification. Once the
product is loaded on the customer's truck, barge, or rail
car, it is then the customer's responsibility.

On the other hand, the customer would prefer quality
control checks to take place as close to the end-use point

as possible. In this case any degradation of the product in transportating and rehandling will be included in the grading analysis.

Problems occur when the actual point of testing is not clearly defined and the producer and customer disagree over whether or not transportation breakage should be included in the product evaluation. This is of particular importance when dealing with large breakwater stone which can be very expensive to produce and even more expensive to replace once it has been installed (Figure 6). Thus it is necessary that all parties agree at what point quality control tests should be run and at what point the specification is to be enforced.

FIG. 6--Failure of 1.8-tonne (2-ton) armor
stone during installation.

SUMMARY

The production and quality control of erosion control stone can be achieved by various techniques. The factors that control the methods employed are primarily the size and quantity of product required, the proposed end use and the quarry at which the stone is being produced. Whatever techniques are chosen, it is the responsibility of all involved to ensure the stone produced will meet specifications and perform as required, resulting in a completed project of which all participants can be proud.

Charles R. Marek

PROPOSED STANDARD FOR SIZES OF STONE FOR EROSION CONTROL

REFERENCE: Marek, C. R., "Proposed Standard for Sizes of Stone for Erosion Control," Rock for Erosion Control, ASTM STP 1177, Charles H. McElroy and David A. Lienhart, Eds., American Society for Testing and Materials, Philadelphia, - 1993.

ABSTRACT: Erosion of embankments on waterways, streams, rivers and lakes, and of fertile agricultural land is a chronic and costly problem. An estimated 4 billion tons of soil erosion occurs annually in the United States. Soil erosion causes water pollution and related environmental concerns, sediment buildup, and loss of valuable real estate. The adverse and costly consequences of soil erosion can be minimized by utilization of effective erosion control methods.

Stone is frequently the first choice of engineers among the erosion control methods that may be considered. However, major deterrents to the consistent supply and use of stone for erosion control applications are (1) the lack of realistic specifications for the grading of quarried stone for this application, and (2) the lack of standardized inspection and testing procedures for establishing compliance with specifications.

ASTM Subcommittee D18.17, Rock for Erosion Control, was established during the late 1970's for the purpose of developing standard specifications and standard test methods for rock for erosion control applications. Since its formation, the subcommittee has developed several test methods for evaluating the quality of stone products to be used for such applications.

The Subcommittee has also developed and is recommending a "Proposed Guide for Specifying Standard Sizes of Stone for

---

*Technical Director, Construction Materials Group, Vulcan Materials Comapny, P. O. Box 530187, Birmingham, AL 35253. Dr. Marek is a member of ASTM Subcommittee D18.17.

Erosion Control." This proposed Guide is presented and
discussed in detail in this paper. Use of the standard
gradings is encouraged for new construction projects, and
standard gradings should be utilized whenever cost-effective
construction will result.

KEYWORDS: erosion control, stone grading, grading specifications, riprap, standardized gradings, filter bedding
stone, quarried stone.

INTRODUCTION

Soil erosion is a chronic and very costly problem.
Erosion of fertile agricultural soil reduces available till-
age areas and crop yields. Erosion of embankments on waterways, streams, rivers, and lakes causes sediment buildup,
water pollution and loss of valuable real estate. Effective
erosion control methods must be utilized whenever possible
to minimize the adverse and costly consequences of soil
erosion.

Soil erosion in the United States has been estimated at
4 billion tons annually. Of this amount, 30 percent is from
natural geologic erosion, 50 percent is from erosion of
agricultural lands, 10 percent is from erosion of forest and
range land, and 10 percent is from erosion at construction
sites. It is estimated that 70 percent of the soil entering
streams and rivers from these sources could be controlled by
use of existing erosion control methods (National Crushed
Stone Association 1982).

Experiments conducted through the years have determined
effective and economical control methods for various areas;
however, few specific engineering guidelines or standards
have resulted which can be used for developing design specifications for a particular protection problem. What has
developed is the use of various materials and methods by the
engineering community. Current control methods include:
a) quarried stone riprap, b) concrete pavement, c) articu-
lated concrete mattresses, d) transverse dikes, e) fences,
f) asphalt concrete mixtures, g) jacks, h) vegetation,
i) gabions, j) erosion-control matting, and k) retaining
walls or bulkheads. This paper is limited to discussion of
one of these methods, quarried stone riprap.

Quarried stone riprap is frequently selected and used
by engineers when stones of sufficient size are available.
Major deterrents to the consistent supply and use of quarried stone riprap are (1) the lack of a standard specifica-

tion for stone grading and quality and (2) the lack of
standardized inspection and testing procedures for establishing compliance. Until a generally accepted grading and
quality specification is established and inspection and
testing procedures are standardized, riprap production will
be a problem for stone producers.

QUARRIED STONE RIPRAP

Quarried stone riprap consists of one or more layers of
stone placed along the bank to be protected. If the slope
of the bank is irregular, the bank is usually graded prior
to stone placement. A porous filter material is placed over
the graded bank to allow seepage of water through the layer
but to prevent erosion of the bank material. The filter
material is covered with larger sized stone to resist the
effects of the moving water.

Quarried stone riprap offers the following general
advantages (U.S. Corps of Engineers 1981b:

a) A quarried stone riprap structure is flexible and
is neither impaired nor weakened by slight movement of the
bank that may result from settlement or other minor adjustments.

b) Local damage loss is easily repaired by placement
of additional rock.

c) Construction usually is not complicated and little,
if any, special equipment or construction practices are
necessary.

d) When stone riprap is exposed to fresh water,
vegetation often grows through the stone layer adding structural value to the material.

e) Appearance is natural and, therefore, acceptable in
all recreational areas.

f) Stone riprap is recoverable and can easily be
stockpiled for future use.

RIPRAP DESIGN OBJECTIVES

Meeting the design objectives necessary to ensure
effective protection requires (1) determination of shape,
size, and mass of the stones in the riprap layer that will
be stable under anticipated hydraulic flow conditions,
(2) use of well-graded bedding material to prevent erosion
of the bank material through the layer, and (3) use of
optimum stone riprap layer and bedding layer thickness.

Empirical relationships have been developed that permit determination of the minimum stone size or mass that will be stable and withstand the maximum hydraulic flow (shear forces) that may occur along the bank to be protected. Experience has shown that the grading of stone used as riprap has little influence on stability when the median mass, W_{50}, is used to characterize stone size (Ahrens 1981). Fifty percent of the stones used in the riprap layer must have "diameters" greater than the computed median diameter, and no more than 50 percent of the stone can have a mass that is less than the computed median mass.

No analytical method has been developed to determine optimum stone shape; the selection of the stone shape is usually a function of subjective experience and what is available. Flat and elongated stones generally are rejected in favor of "block-type-stones" because the block-type fit together better. Lift and drag forces on flat stones and drag forces on elongated stones are greater in proportion to the stone mass than are forces on the more desirable angular and blocky shapes. Stones with a maximum dimension greater than three times their minimum dimension should not be considered. Sharp edges are preferred over rounded edges for increased stability.

The grading of stone used in a stone erosion control structure influences structural performance. Poor grading encourages failure of the structure because oversize stones may preclude mutual mechanical support among individual stones. If the grading of the stone is such that movement of underlying material through the stone layer is likely, a filter of sand, crushed rock, gravel, or synthetic cloth should be placed under the quarried stone layer. The ideal riprap structure design provides a gradual reduction in stone size until stone in the layer blends with the size of the natural bed material; however, this design is seldom economically justified.

CURRENT SIZE SPECIFICATIONS

In 1967 the National Crushed Stone Association published a two-part document (National Crushed Stone Association 1967) pertaining to stone for coastal structures. The following statements were made in this document:

"In the fall of 1966, the U.S. Army Coastal Engineering Research Center released Technical Report No. 4 (entitled) 'Shore Protection, Planning and Design,' which is fundamentally a textbook on the subject of coastal structures. The report, a worthwhile addition to any stone producer's library, develops and illustrates the principles involved in the design and construction of coastal structures, but offers little help to the individual faced with the production and supply of stone necessary for such projects.

"From the number of inquiries received by the NCSA office, the question of riprap specifications, riprap production, and control techniques is apparently a pressing one."

Twenty-five years have passed since the NCSA document was published. During these years, efforts have been made to standardize specifications for riprap and to develop and standardize methods for establishing compliance with specifications. Unfortunately, standardization has not been achieved to date.

Specifications for quarried stone riprap vary with specifying agencies, and often vary from project to project even when specified by the same agency and for similar conditions. In recent years, the number of different gradings specified for quarried stone riprap has been increasing. This situation exists because empirical formulas are used for design of riprap structures, and these formulas establish the size and mass of stone required to impart necessary stability to the structure being designed. Often differences in grading specifications are relatively minor, but are sufficient to cause production problems and preclude stockpiling of riprap products at most stone production facilities.

Because many different gradings may be required from a given quarry, production costs are "high" due to the need to change the equipment used to produce riprap to each special grading. For the production of stone riprap to be cost-effective, an order of at least one week's production is normally required (U.S. Corps of Engineers 1981a). This means that producers must either (1) stockpile each specific grading of quarried riprap stone that may be required, or (2) incur increased production costs (with increased cost to purchaser), or (3) elect not to produce the specified product. Many small orders of quarried stone riprap receive little or no interest because the producer does not have the stone stockpiled or does not expect to receive another order for the same grading of stone in the immediate future. As a consequence, product availability is reduced and construction costs are increased. Standardization of stone riprap gradings will increase product availability and should decrease construction costs.

PROPOSED STANDARD SIZES

A subcommittee in the American Society for Testing and Materials (ASTM Subcommittee D18.17) was established in the late 1970's for the purpose of developing standard specifications and standard test methods for rock for erosion control applications. This Subcommittee has been active and has developed a "Proposed Guide for Specifying Standard Sizes of Stone for Erosion Control." The proposed Guide was

TABLE 1--Standard sizes of graded stone

Approx. size[a]		Particle mass[c]		Percent lighter than the mass specified[d]						
(cm)	(in.)	(kg)[b]	(lb)	R-3000	R-1500	R-700	R-300	R-150	R-60	R-20
107	42	3 400	7 500	100	...	...	...	...	...	...
81	32	1 400	3 000	50-100	100	...	...	...	...	...
63	25	680	1 500	15-50	50-100	...	...	...	...	...
56	22	450	1 000	...	...	100	...	...	...	...
48	19	320	700	...	15-50	50-100	100	...	...	...
43	17	130	500	0-15	...	...	...	...	...	...
38	15	140	300	...	...	15-50	50-100	100	...	...
36	14	110	250	...	0-15	...	...	...	...	...
30	12	68	150	...	...	...	15-50	50-100	100	...
23	9	27	60	...	...	0-15	...	15-50	50-100	...
20	8	20	45	...	...	...	0-15	...	...	100
18	7	14	30	...	...	...	...	...	15-50	...
15	6	9.1	20	...	...	...	...	0-15	...	50-100
10	4	4.5	10	...	...	...	...	...	0-15	15-50
8	3	0.9	2	...	...	...	...	...	...	0-15

a Maximum dimension of individual stone. The ratio of the maximum to minimum dimension of the stone shall not exceed 3. (1 in. = 2.54 cm)
b Rounded to two figures from conversion of U.S. Customary Units.
c Calculated assuming a cube with a specific gravity of 2.65. Appropriate adjustments in the table values should be made for stone particles having specific gravity values other than 2.65. (1 kg = 2.2 lb)
d Established by determining the mass of the individual stone particles.

Table 1 after reference Marek, 1982.

TABLE 2--Standard sizes of filter bedding stone

Size identification:		FS-3	FS-2	FS-1
Sieve size (Sq. opening)[a]				
SI unit (mm)	Alternate Units	Percent smaller than the size indicated[b]		
165	(6-1/2 in.)	100	...	...
114	(4-1/2 in.)	85-100	...	...
63	(2-1/2 in.)	15-50	...	...
50	(2 in.)	...	100	...
19.0	(3/4 in.)	...	85-100	...
9.5	(3/8 in.)	...	...	100
4.75	(No.4)	...	15-50	85-100
1.18	(No.16)	0-15	...	...
0.60	(No.30)	...	...	15-50
0.15	(No.100)	...	0-15	0-15

a As specified in ASTM E 11.
b Determined in accordance with ASTM C 136.

Table 2 after reference Marek, 1982.

initially published in the September/December issue (1982)
of the ASTM *Geotechnical Testing Journal* (Marek 1982). The
latest revision is now in the ASTM balloting process.

The proposed Guide contains new specifications pertaining to sizes (gradings) of stone products for erosion control applications that have heretofore not been standardized. A need presently exists for such standardization to
(1) offer the designer, the consumer, and the producer a
common reference in connection with sizing quarried stone
for use in erosion control structures and (2) provide the
most economical products in the volume required.

The standard sizes of stone for erosion control as set
forth in the ASTM Proposed Guide are separated into two
categories:

(1) Graded stone sizes [Table 1].

(2) Filter bedding stone sizes [Table 2].

Use of standard gradings should be encouraged for all major
projects, and standard gradings should be utilized whenever
cost-effective.

SUMMARY

The designer establishes the size of graded quarry
stone required for the project using acceptable design
criteria. Consideration should then be given to using one
of the proposed graded stone sizes contained in Table 1.
This will be primarily an economic consideration. Selection
of the next larger "standard" size may result in an overdesigned layer having increased reliability, but will also
require an increased volume of stone. The cost-effective-
ness of using one of the "standard" gradings versus a "nonstandard" grading should be evaluated, and a standard grading should be utilized whenever feasible. Good engineering
practice allows designers to accept an "overdesigned" job in
order to decrease the costs associated with producing,
transporting, stockpiling, and placing nonstandard gradings
of graded quarry stone and thereby reduce the overall
project cost.

In a properly designed riprap layer, the graded stone
particles should be contained reasonably well within the
layer thickness. Oversize stones, even in isolated locations, can cause failure of the graded stone layer by pre-
cluding mutual support between individual stones, providing
large voids that expose filter or bedding materials, and
creating excessive local turbulence that removes smaller
stones.

Standardized gradings will serve user needs for most design situations. There may be cases, however, where a "nonstandard" grading will be required. Nonstandard gradings should be considered and specified only when they can be justified on a cost-effective basis.

REFERENCES

Ahrens, John P., December 1981, "Design of Riprap Revetments for Protection Against Wave Attack, U.S. Corps of Engineers, Technical Paper No. 81-5, 33 pp.

Marek, C. R., Sept/Dec 1982, "Proposed Standard Specifications for Standard Sizes of Quarried Stone for Erosion Control," ASTM Geotechnical Testing Journal, Vol. 5, No. 3/4, pp. 93-95.

National Crushed Stone Association, September 1967, "Stone for Coastal Structures," Useful Information Series, Part I - General Principles - August 1967 and Part II - Riprap Specifications and Production Tips, (Washington, DC).

National Crushed Stone Association, May 1982, "Quarried Stone for Erosion and Sediment Control," 33 pp, (Washington, DC).

U.S. Corps of Engineers, November 1981a, "Report on Standardization of Riprap Gradation," Lower Mississippi Valley Division, 13 pp.

U.S. Corps of Engineers, December 1981, "The Streambank Erosion Control Evaluation and Demonstration Act of 1974, Section 32, Public Law 93-251," Final Report to Congress.

Richard J. Lutton[1] and G. Sam Wong[2]

RIPRAP QUALITY CRITERIA IN STANDARD SPECIFICATIONS AND ENGINEERING
GUIDANCE

REFERENCE: Lutton, R. J. and Wong, G. S., **"Riprap Quality Criteria
in Standard Specifications and Engineering Guidance,"** Rock for Erosion
Control, ASTM STP 1177, Charles H. McElroy and David A. Lienhart, Eds.,
American Society for Testing and Materials, Philadelphia, 1993.

ABSTRACT: Several tests and requirements have been used routinely in
the past 30 years for durability and quality of stone for use as riprap
and armor. Among the tests are absorption, unit weight, abrasion,
sulfate soundness, and freezing-thawing. The tests all give index
values reflecting durability only indirectly. Some Federal agency
guidance and the standard specifications of state highway departments
are reviewed here to translate the index values into criteria for
evaluating stone quality as far as suitability or unsuitability.

KEYWORDS: stone, erosion control, riprap, durability, specifications,
guidance, quality, tests

INTRODUCTION

The purpose of this review is to characterize riprap stone that is
acceptable for erosion protection in the context of standard
specifications and methods required routinely for many years.

The review focuses primarily on materials classified as riprap in
sizes mostly less than 225 kg (500 lb). Materials composed of very
large stone (up to 13000 kg) such as used on jetties and harbor
structures are mostly excluded except in making a comparison between the
performance of these very large-stone materials and the performance of
riprap. Information on performance and quality of very large stone
materials has been presented elsewhere (Lutton 1992; Lutton and Erickson
1992).

---

[1]Geologist, Geotechnical Laboratory, USAE Waterways Experiment
Station, Vicksburg, MS 39180-6199.

[2]Petrographer, Structures Laboratory, USAE Waterways Experiment
Station, Vicksburg, MS 39180-6199.

The economic importance of riprap and larger stone materials has been emphasized previously (Lutton 1990) by reference to published and unpublished data of the Bureau of Mines indicating U.S. production of about 25 million short tons (22.7 MMg) in 1985.

The background for this review includes:

- Studies on practices and quality problems of the Corps of Engineers (CE) intermittently from 1974 through 1991 by Lutton (Corps of Engineers 1990; Lutton 1992; Lutton and Erickson 1992; Lutton 1990; Lutton et al. 1981).

- Material testing and petrographic studies by Wong and others. The Waterways Experiment Station (WES) has been testing riprap stone since the 1960s and was instrumental in formalizing tests and petrographic methods used throughout the CE.

A companion paper in this volume by Wong and Lutton reviews petrographic methods and criteria used currently at WES and suitable for use elsewhere as well.

STATE SPECIFICATIONS

Criteria for riprap or equivalent features for erosion control are found in standard specifications of highway or transportation departments in each of the 50 states. These separate standards, each reflecting the experience and expertise of a large and active engineering organization, constitute a substantive file. Together the standards provide a synopsis of common criteria on the one hand and on the other hand indicate some departures in the nature of special testing. Table 1 lists the standard specifications consulted for each state with the date of the particular edition. Where supplemental specifications or amendments were available, they were checked for revisions made subsequently. Almost all of the criteria are found under sections dealing with materials. Table 2 summarizes the results of this review. Ten criteria were distinguished.[3]

Six of the ten criteria are in terms of maximum or minimum values in testing the stone materials by standard method. Five of the six tests are primarily methods for evaluating aggregate for concrete.

- Soundness is approximated using Standard Method of Test for Soundness of Aggregate by Use of Sodium Sulfate or Magnesium Sulfate (AASHTO T 104/ASTM C 88) (AASHTO 1990; ASTM 1991) or variants or similar alternative tests preferred by some states. Sodium sulfate is used more commonly than magnesium sulfate.

---

[3]Criteria unique to only one state's specifications, such as California's durability index and associated test, are not included.

- Resistance to abrasion or wear is approximated using Standard Method of Test for Resistance to Abrasion of Small Size Coarse Aggregate by Use of the Los Angeles Machine (AASHTO T 96/ ASTM C 131) or variants or similar alternative tests preferred by some states.

- Absorption is determined using AASHTO Standard Method of Test for Specific Gravity and Absorption of Coarse Aggregate (T 85) or variants or similar alternative tests preferred by some states.

TABLE 1--Standard specifications reviewed for riprap.

State	Edition		State	Edition	
Alabama	1985		Montana	1987	S.S. 01-88
Alaska	1988		Nebraska	1985	
Arizona	1987		Nevada	1986	
Arkansas	1988		New Hampshire	1983	S. 84
California	1988		New Jersey	1983	
Colorado	1986		New Mexico	1984	
Connecticut	1988	S.S. 7-89	New York	1985	Add. 06-88
Delaware	1985		North Carolina	1984	
Florida	1986		North Dakota	1986	S. 11-88
Georgia	1983	S.S. 86	Ohio	1987	
Hawaii	1985		Oklahoma	1976	S. 84
Idaho	1983	S.S. 10-87	Oregon	1984	
Illinois	1983	S.S. 10-86	Pennsylvania	1987	
Indiana	1985	S.S. 01-88	Rhode Island	1971	Rev. 12-91
Iowa	1984		South Carolina	1986	
Kansas	1980		South Dakota	1985	
Kentucky	1988		Tennessee	1981	Rev. 07-87
Louisiana	1982		Texas	1982	
Maine	1988		Utah	1979	Add. 02-85
Maryland	1982	S. 01-88	Vermont	1986	
Massachusetts	1988		Virginia	1987	
Michigan	1984		Washington	1988	
Minnesota	1983	S.S. 05-87	West Virginia	1986	S.S. 01-88
Mississippi	1976	S.P. 90	Wisconsin	1981	S.S. 05-88
Missouri	1986		Wyoming	1987	

S.S. - Supplemental Specifications, S.P. - Special Provision, Add. - Addendum, Rev. - Revision, S. - Supplement

- Unit weight is determined using AASHTO T 85 or variants or
  similar alternative tests preferred by some states.

- Freezing-thawing vulnerability testing is specified only in
  Iowa, Nebraska, Indiana, and New York, each with a separate
  method performed on small stone pieces.

The sixth quantitative parameter is the size gradation of riprap
material. Conformance to the specified gradation is usually estimated
visually, although a few state specifications indicate that checks may
be required by actual measurement, either stone by stone or by weighing
size classes.

TABLE 2--State specifications for characterizing riprap.

Character	Number (and percent) of states out of 50
Soundness	14 (28)
Abrasion	15 (30)
Absorption	7 (14)
Unit Weight	20 (40)
Freezing-Thawing Resistance	4 (8)
Stone Gradation	46 (92)
Narrative Description	42 (84)
Geology/Petrography	35 (70)
Approved Source	10 (20)
Stone Shape	36 (72)

The remaining four of the ten state criteria distinguished in this
review are qualitative and rely at least partly on scientific or
engineering judgment and experience to be effective.

- Geologic and petrographic aspects of riprap materials are
  mentioned critically in 70 percent of the specifications, e.g.
  clay content, stratification, and specific rock types. North
  Dakota prohibits local stone of sedimentary origin and Virginia
  prohibits the use of sedimentary rock in features exposed to
  the action of sea water. A formalized petrographic examination
  such as according to Standard Guide for Petrographic
  Examination of Aggregates for Concrete (ASTM C 295) was
  required in only one state specification. From this, one might
  infer that the experience of the geologist or petrographer is
  most critical. In the exception (Georgia) a favorable
  appraisal in the petrographic analysis can override
  unsatisfactory test results.

- State specifications almost always first define the desirable characteristics of riprap material in simple, general terms implying performance. For example the requirements are sometimes for strong, durable stone, adequate to withstand the effects of weathering and to perform the intended function.

- Approved or acceptable material or source of material, where distinguished, was considered in this review to be essentially synonymous with a satisfactory service record of the same source or rock formation on a previous project.

- Most specifications also require a favorable shape for individual stones. In some the dimensional ratios are given while in others the stone is required to be rectangular, simply angular, or with at least a minimum number of fracture faces.

CORPS OF ENGINEERS AND BUREAU OF RECLAMATION PRACTICES

Table 3 summarizes past practices for riprap in two large Federal organizations constructing slope protection for embankments, dams,

TABLE 3--Methods for characterizing riprap in federal guidance.

Character	Corps of Engineers[a]	Bureau of Reclamation[b]
Soundness	X	X
Abrasion	X	X
Absorption	X	X
Unit Weight	X	X
Freezing-Thawing Resistance	X	X
Stone Gradation	X	X
Narrative Description	X	X
Geology/Petrography	X	X
Approved Source	X	X
Stone Shape	X	X

[a] Engineer Manual 1110-2-2302 (Corps of Engineers 1990)
[b] Designation E-39 Investigations for Rock Sources for Riprap (Water and Power Resources Service 1974). E-39 is identified as still applicable in the third edition of the Earth Manual published in 1990.

irrigation canals, levees, channels, and shorelines. Each agency uses all of the ten previously distinguished means of describing riprap material at least selectively. In this context these two Federal organizations seem to have been more demanding, but factors other than thoroughness probably account for the difference. These factors are discussed below under CRITERIA FOR EVALUATION. CE practice and guidance (Corps of Engineers 1990) are emphasized in the following review:

- Petrographic Examination. The composition and homogeneity of
  samples and their general physical condition are characterized
  with emphasis on flaws that may be found in large stones.
  Accordingly, samples are selected to include flaws common to
  large stone. Among special methods used for studying large
  stones are polishing, etching, and staining of cut slabs.
  Serious defects identifiable in these ways are platiness,
  shaliness, slabbiness, and a tendency to slake. Potential for
  such defects may be present in the form of clay seams, bedding,
  blasting fractures, joints, rounded or planar surfaces,
  nodules, and indications of weathering or chemical alteration.
  High-quality stone sometimes exhibits an interlocking fabric
  and absence of bedding. A useful technique is to wipe the
  rough or cut stone with a wet cloth to emphasize defects.

- Specific Gravity. Care should be exercised in using specific
  gravity to characterize stone since only the solid components
  (mineralogical) are considered in true specific gravity.
  However, the terms "apparent specific gravity" and "bulk
  specific gravity (saturated, surface-dry basis)," adapted from
  aggregate testing in Standard Test Method for Specific Gravity
  and Absorption of Coarse Aggregate (ASTM C 127), are entrenched
  in past experience, and any departure, regardless of its
  sensibility, may introduce ambiguity. Carefully defining and
  limiting such terms in the specifications is essential to
  avoiding ambiguity. A more useful parameter sometimes is dry
  unit weight in which the important parameter porosity is
  included, provided that specific gravity of solids can be
  estimated confidently.

- Absorption. A portion of rock porosity functions to draw water
  in from the surface by absorption. Absorption of water is a
  common precursor of stone deterioration, and the absorption
  test (ASTM C 127) is particularly useful for revealing
  vulnerability.

- Sulfate Soundness. Standardized testing follows ASTM C 88,
  again a method developed for evaluating aggregate. Samples
  soaked in magnesium sulfate solution[4] will break apart when the
  solution invades weak planes or cracks and then crystallizes
  upon heating and drying. A major shortcoming of this test for
  large stone is that the test specimens are broken from the
  large stone to a weight of approximately 100 g each. The
  breakage and segregation tend to eliminate weak areas when

---

[4]The Bureau of Reclamation uses sodium sulfate solution (Water and
Power Resources Service 1974) which tends to be less severe in its
effect.

preparing the sample, and test results tend to be too
favorable. The test is usually meaningful for sedimentary
rocks when augmented by an absorption or abrasion test, except
for some sandstones.

- Ethylene Glycol Soundness. Standardized testing follows the CE
  Method of Testing Stone for Expansive Breakdown on Soaking in
  Ethylene Glycol (CRD-C 148) (Waterways Experiment Station
  1949). This method is used to detect the presence of swelling
  clay minerals and provides an indication of the severity of
  deterioration of the stone to be expected in service. Ethylene
  glycol enters the clay mineral structure and causes rapid
  expansion. The test has been particularly useful in
  distinguishing questionable varieties among altered basaltic
  rocks.

- Abrasion Resistance. Standard Test Method for Resistance to
  Degradation of Large-Size Coarse Aggregate by Abrasion and
  Impact in the Los Angeles Machine (ASTM C 535) approximates the
  resistance of stone to abrasion and battering and also provides
  an index of toughness, durability, and abundance of incipient
  cracks. The significance of the test for large stone is
  indefinite since individual test pieces are limited to about
  100 g in weight. Weaknesses along widely spaced surfaces are
  missed in this test. The test is sometimes effective for
  evaluating metamorphic rock, particularly when supported by
  absorption and sulfate soundness tests.

- Freezing-Thawing Resistance. The standard CE Method of Testing
  Stone for Resistance to Freezing and Thawing (CRD C 144)
  (Waterways Experiment Station 1949) was developed at WES in the
  mid 1960s. Weathering effects of a cold environment are
  simulated by inducing numerous cycles of freezing and thawing
  to a large stone slab through a bath of water and alcohol. The
  number of cycles to which the specimen is subjected and the
  overall interpretation of the results are usually determined on
  a regional or laboratory basis. The number of cycles commonly
  exceeds 10, occasionally going to 50 or more.

- Wetting-Drying Resistance. Testing large stone for wetting and
  drying effects generally follows laboratory-level guidance
  since no standard method is used CE-wide. A standard method
  suitable for testing large stone has been proposed (Lutton
  et al. 1981). No generally applicable experiences are
  available correlating quantitative test results and stone
  service in place. Considerable judgment has to be exercised
  even in descriptions of scaling and flaking, random cracking,
  and slabbing along bedding and similar fabric. Photographs are
  helpful in characterizing the rock and its behavior in regard
  to deterioration.

CRITERIA FOR EVALUATION

Test Values

Table 4 presents test values and ranges of values found in the state standard specifications and CE guidance discussed in the previous paragraphs. The CE guidance is broader and more demanding of stone quality. The conspicuously demanding CE guidance in comparison to state specifications is explained by some or all of the following:

- CE criteria in the table are intended as expected properties of only the best stone. Values ranging below those listed are common among suitable stone. For example, the unit weight of stone successfully used has commonly ranged down to 2.24 g/cm^3 (140 lb/ft^3) and absorption values up to 2 percent are also common.

- The CE guidance is biased in application toward very large, high-cost stone materials such as for armor. In recent experience (Lutton 1992) most CE problems have been in northern regions, particularly around the Great Lakes. Cracking and durability problems there center on the use of marginal rock materials for large-size armor stone exposed to a severely cold winter environment. A conservative approach emphasizing the use of high quality stone is taken there and has been extended more generally elsewhere.

- Breakwaters, dam embankments, and other hydraulic structures commonly present environments in which stresses on rock are more severe than they are along road embankments and ditches typical of highway construction. Also the consequences of failure are often more serious.

- It is not uncommon in highway construction to select locally available stone despite somewhat marginal quality and in the process to accept higher maintenance costs in exchange for lower construction costs.

Non-Test Characteristics

Narrative descriptions and geologic or petrographic evaluations constitute the most common means of characterizing stone quality. Eighty-four percent of state specifications characterize suitable stone in terms such as "durable," "hard," and "dense." Potential problems in enforcing this type specification are well known. Seventy percent of state specifications place geologic and petrographic constraints on the materials such as the suitability or unsuitability of certain rock types or mineral components and the absence of seams, cracks, and other geological structures. Again the emphasis on high quality for very large stone materials accounts for the routine inclusion of a formal petrographic examination in most CE stone evaluations.

TABLE 4--Test criteria for evaluating riprap.

	Typical Criteria	
Characteristic	State Specification (Range)	CE Guidance
Soundness, % loss		
Sodium Sulfate (max)	12-20	--
Magnesium Sulfate (max)	10-20	5
Ethylene Glycol (max)	--	0
Abrasion, % loss (max)	40-60	20
Absorption, % (max)	2-6	1
Stone Density		
Specific Gravity (min)	2.3-2.5	--
Unit Weight, g/cm^3 (min)	2.24-2.64[a]	2.56[b]
Freezing-Thawing, % loss		
Coarse Aggregate (max)	10-14	--
	(16-25 cycles)	
Stone Slab (max)	--	10
		(12 cycles)
Wetting-Drying, % loss		
Stone Slab (max)	--	0
		(35 cycles)

[a] 140-165 lb/ft^3
[b] 160 lb/ft^3

SUMMARY AND CONCLUSIONS

Certain tests that have been more or less standardized are widely accepted and used for evaluating stone as riprap. Most of these tests are adapted or have evolved from those used extensively for concrete aggregate. The tests explore in a simplified way the resistance of the rock to abrasion, freeze-thaw, and wet-dry environmental attack. The wide acceptance provides a useful frame of reference for comparing test results and applicability.

The most consistent application and usefulness of formalized laboratory testing have been in evaluating very large stone materials such as for armor on coastal and harbor structures. Application there follows from the high cost and importance of these major structures. Test criteria for riprap are found in only some of the state specifications and then on a selective basis. Previous satisfactory service of the material in question is more definitive.

Much of the selection of the relatively small stone needed for riprap in highway construction is made on the basis of:

• Narrative descriptions of material quality.

• Use of established sources with satisfactory service records.

• Exercise of technical judgment based on previous experience.

ACKNOWLEDGMENT

Observations presented herein are an outgrowth of research conducted under the Civil Works Guidance Update Program of the United States Army Corps of Engineers by the Waterways Experiment Station. Permission was granted by the Chief of Engineers to publish this information.

REFERENCES

American Association of State Highway and Transportation Officials, 1990, Standard Specifications for Transportation Materials and Methods of Sampling and Testing.

American Society for Testing and Materials, 1991, Annual Book of ASTM Standards, Section 04, Construction, (updated annually).

Corps of Engineers, 1990, "Construction with Large Stone," Engineer Manual 1110-2-2302, Department of the Army, 58 pp.

Lutton, R. J., 1992, "U.S. Experience with Armor-Stone Quality and Performance," in Durability of Stone for Rubble Mound Breakwaters, American Society of Civil Engineers, pp. 40-55.

Lutton, R. J., 1990, "Material Characteristics of Large Stone in American Construction Practice." Vol. 286, Transactions of Society of Mining, Metallurgy, and Exploration, pp 1850-1855.

Lutton, R. J. and Erickson, R. L., 1992, "Problems with Armor-Stone Quality on Lakes Michigan, Huron, and Erie," in Durability of Stone for Rubble Mound Breakwaters, American Society of Civil Engineers, pp 115-136.

Lutton, R. J., Houston, B. J., and Warriner, J. B., 1981, "Evaluation of Quality and Performance of Stone as Riprap and Armor," TR GL-81-8, U.S. Army Engineer Waterways Experiment Station, 117 pp.

Water and Power Resources Service, 1974, Earth Manual, Second Edition, Bureau of Reclamation.

Waterways Experiment Station, 1949, Handbook for Concrete and Cement, Corps of Engineers, Department of the Army, (updated periodically).